Move Towards Zero Hunger

Basanta Kumara Behera
Pramod Kumar Rout
Shyambhavee Behera

Move Towards Zero Hunger

Springer

Basanta Kumara Behera
Advanced Centre for Biotechnology
Maharshi Dayanand University
Rohtak, Haryana, India

Shyambhavee Behera
Department of Community Medicine,
University of Delhi
University College of Medical Sciences,
GTB Hospital
New Delhi, India

Pramod Kumar Rout
Genetics and Breeding Division
ICAR-Central Institute for Research
on Goats, Makhdoom
Mathura, Uttar Pradesh, India

ISBN 978-981-32-9799-9 ISBN 978-981-32-9800-2 (eBook)
https://doi.org/10.1007/978-981-32-9800-2

This Springer imprint is published by the registered company Springer Nature Singapore Pte Ltd.
The registered company address is: 152 Beach Road, #21-01/04 Gateway East, Singapore 189721, Singapore

Preface

The inhabitable world is gradually losing social integrity, and as a result, an existence of huge gap between rich and poor has brought brainstorming impact to resolve such problem in an amicable way under the provision of minimum budget. Some parts of global geographic locations, rich in modern agriculture technologies, are in dilemma to store the overproduced food grains, and even seasonal fruits and vegetable, and are forced to destroy the overproduced agricultural products under the constraint of poor logistic system, food warehouses and management of micro-economy system. On the other side, millions of people in extreme rural captured area are leading stressful life with hunger and poverty. This piece of book gives clues to resolve the problems of food security and poverty in rural areas so that minimum social justice will be accessed to the rural people deprived of regular food and shelter. This book also projects how to develop sustainable foundation in extreme rural location with the help of indigenous resources to confront with problems like hunger, malnutrition and chronic health problem by structural transformation of prevailing extremely poor social economic transition and awfully deprived life style with scarcity of minimum food and shelter and bring them in sustainable form for better life for tomorrow.

Virtually, Nobel concept of zero hunger programme was initiated by President Lula in 2003 for hunger eradication and social inclusion, linking macro-economic, social and productive polices. After a decade and so, zero hunger challenge was launched by United Nations Secretary-General Ban Ki-moon in 2012. As reported in the Global Hunger Index 2017, the world has made progress in reducing hunger since 2000; however, the progress has been uneven, and not fulfilling the claims, earlier. Recent report says that several countries have higher hunger level in 2017 than they did in 2008, indicating that progress in reducing hunger over recent decades is reversing more recently in some places. Due to the scarcity of data bank and reference books, it has been a herculean task to resolve the problem of extreme poverty and go ahead with zero hunger in a substantial manner. So, the main reason for designing this piece of manuscript is to provide ready reference to the sciences those who are having social volunteers; NGO; international organisations like United Nations Rome-based agencies (RBAs), the Food and Agriculture Organization of the United Nations (FAO), the International Fund for Agricultural Development (IFAD) and World Food Programme (WFP); policymakers;

bureaucratic politicians; and students studying sociology, social sciences, agriculture, public health and veterinary inquisitive mind to know more about zero hunger.

The authors are highly thankful to Mrs. Asha Sharma, Katyayanee, Gopal, Ms. Mahima Verma (Postdoctoral Fellow) and Mrs. Rashmirekha for their immense help in computer graphic design, model preparation and data compiling while preparing and giving shape to the manuscript. The kind cooperation of the publisher is highly acknowledged. We sincerely acknowledge the help of "Consortium of Scientists and Sociologists for Social Inclusion Development (CSSSID)," while computerizing data mine on rural women's empowerment.

Rohtak, India — Basanta Kumara Behera
Mathura, India — Pramod Kumar Rout
New Delhi, India — Shyambhavee Behera

Contents

About the Authors

Basanta Kumara Behera was a Professor of Biotechnology at three distinguished Indian universities, and has been involved in postgraduate teaching and research on the topics related to bio-energy management and biomass processing technology since 1978. In 2009 he joined a MNS company as adviser for speciality chemical production and drug design using microbial process technologies. Prof. Behera is associated with leading national and international companies as a technical adviser on the production of biopharmaceuticals under cGMP norms. He has authored various books published by CRC Press USA and Springer Verlag, Germany, and is also an accomplished writer, writing in many forms like free verse, haiku, Senryu, Sonnet, Tanka, and quatrain. He publishes English poems in both national and international magazines.

Pramod Kumar Rout is working as a Principal Scientist at ICAR-CIRG. He completed his doctoral degree at the Indian Veterinary Research Institute, Izatnagar and worked as a postdoctoral fellow at the National Institute of Immunology, New Delhi and at the Genetics and Genomics Division, Roslin Institute, Edinburgh, UK. He has also worked as a Visiting Scientist at the University of Arizona, USA and he coordinates an All India research project on goat improvement in different agro climatic zones in India. His research interests include the genetic diversity of livestock, conservation of animal genetic resources and management, genetics of disease resistance, genetics of heat stress regulation in livestock and milk genomics. He has also worked in the area of conservation of endangered goats in their natural habitat and on various developmental projects with farmers in remote corners of India.

Shyambhavee Behera is a medical graduate from Lady Hardinge Medical College, New Delhi, and holds an MD degree from University College of Medical Science, New Delhi. She works in the field of non-communicable disease, immunization, maternal and child health, epidemiology and health administration to promote sustainable life styles and improve health. Currently she is working as Senior Resident, Department of Community Medicine, Atal Bihari Vajpayee Institute of Medical Sciences and Dr, Ram Manohar Lohia Hospital, New Delhi, India. She was the co-author of book series "New Paradigms of Living Systems" SpringerNature, Singapore Pte. Ltd.

Abbreviations

AFAD	Académie de Football Amadou Diallo de Djékanou
AIDS	Acquired immune deficiency syndrome
AFOLU	Agriculture, Forestry and Other Land Use
ACOG	American College of Obstetricians and Gynaecologists
ANGELS	Antenatal and Neonatal Guidelines, Education and Learning System
AGP	Antimicrobial growth promotants
AMR	Antimicrobial-resistant bacteria
CAPEX	Capital expenditures
CDC	Centers for Disease Control and Prevention
CMS	Centers for Medicare and Medical Services
CHIP	Children's Health Insurance Program
CFC	Chlorofluorocarbons
CSA	Climate Smart Agriculture
CEDAW	Committee on the Elimination of Discrimination Against Women
CFL	Compact fluorescent lamp
CAGR	Compound annual growth rate
CPOE	Computerised physician order entry
CPOM	Computerised provider order management
CCT	Conditional cash transfer
DGC	Dairy goat cooperative
DESA	Department of Economic and Social Affairs
DREI	Derisking Renewable Energy Investment
ECHO	Educational Concerns for Hunger Organization
EAEMP	European Agency of the Evaluation of Medical Products
EC	European Commission
FOI	Fairtrade Labelling Organizations International
FF	Family farming
FCE	Feed conversion efficiency
FGM	Female genital mutilation
FFVs	Flexible Fuel Vehicles
FANRPAN	Food Agriculture and Natural Resources Policy Analysis Network
FAOSTATS	Food and Agriculture Organization Corporate Statistical Database
FAO	Food and Agriculture Organization
FDA	Food and Drug Administration

GHI	Global Hunger Index
GLP	Good Laboratory Practices
GMP	Good Manufacturing Practices
GHG	Greenhouse gas
HACCP	Hazard analysis and critical control points
HHS	Health and Human Services
HCFA	Health Care Financing Administration
HAI	Healthcare-associated infection
HIT	Health information technologies
HRSA	Health Resources and Services Administration
HLPF	High-Level Political Forum on Sustainable Development
HIV	Human immunodeficiency virus
IgG	Immunoglobulin
IMR	Infant mortality rate
IPC	Infection prevention and control
ICT	Information and communication technologies
ISFM	Integrated soil fertility management
IEA	International Energy Agency
IFPRI	International Food Policy Research Institute
IFAD	International Fund for Agricultural Development
IHP+	International Health Partnership
ILO	International Labour Organization
IMF	International Monetary Fund
IRENA	International Renewable Energy Agency
JMP	Joint Monitoring Programme
Kg	Kilogram
LDCs	Least developing countries
LED	Light-emitting diode
LPG	Liquid petroleum gas
LBW	Low birth weight
MAPS	Mainstreaming, Acceleration and Policy Support
MHM	Menstrual hygiene management
CH_4	Methane
MDGs	Millennium Development Goals
MSD	Millennium Sustainable Development
ng/L	Nanogram per litre
NCAs	Natural Capital Accounts
N	Nitrogen
NGO	Nongovernment organisation
OMB	Office of Management and Budget
ORHP	Office of Rural Health Policy
OECD	Organization for Economic Cooperation and Development
ODS	Ozone-Depleting Substances
P	Phosphorous
PMJDY	Pradhan Mantri Jan Dhan Yojana

CST	Prevalence of stunting in children under 5 years old (in %)
CWA	Prevalence of wasting in children under 5 years old (in %)
PHC	Primary healthcare
CM	Proportion of children dying before the age of 5 (in %)
PUN	Proportion of the population that is undernourished (in %)
RBAs	Rome-based agencies
RHAP	Rural Health Advocacy Project
RHRPRC	Rural Health Reform Policy Research Center
RHWA	Rural Health Workforce Australia
SHGs	Self-help groups
SRHR	Sexual and Reproductive Health and Right
SIWI	Stockholm International Water Institute
SAG	Strategic Advisory Group
SCM	Supply Chain Management
SDGs	Sustainable Development Goals
IFAD	The International Fund for Agricultural Development
UNICEF	The UN Children's Fund
TTI	Think Tank index Report
TTCSP	Think Tanks and Societies Program
ESCR	UN Committee on Economic, Social and Cultural Rights
ECOSOC	UN Economic and Social Council
UNHCR	UN High Commissioner for Refugees
DFID	UK Department for International Development
UK	United Kingdom
UN	United Nations
UNCDF	United Nations Capital Development Fund
UNCF	United Negro College Fund
UNICEF	United Nations Children's Fund
UNCTAD	United Nations Conference on Trade and Development
UNDAEF	United Nations Development Assistance Frameworks
UNDG	United Nations Development Group
UNDP	United Nations Development Programme
UNEP	United Nations Environment
UNICEF	United Nations International Children's Emergency Fund
UNISDR	United Nations Office for Disaster Risk Reduction
RBAs	United Nations Rome-based agencies
US	United States
USAID	United States Agency for International Development
USA	United States of America
UHC	Universal Healthcare
UHC	Universal health coverage
USDA	US Department of Agriculture
VAD	Vitamin A Deficiency
VNRs	Voluntary national reviews
WASH	Waste management and environmental cleaning

WASH	Water, Sanitation and Hygiene
WFP	World Food Programme
WFS	World Food Summit
WHA	World Health Assembly
WHO	World Health Organization
ZHC	Zero Hunger Challenge

Move Towards Zero Hunger

1

1.1 Envisioning Zero Hunger

The habitable world is gradually losing social integrity, and as a result an existence of huge gap between rich and poor has brought brain storming impact to resolve such problem in an amicable way under the provision of minimum budget. Some parts of global geographic locations, rich in modern agriculture technologies, are in dilemma to store the overproduced food grains, and even seasonal fruits and vegetable, and are forced to destroy the overproduced agricultural products under the constraint of poor logistic system, food warehouses and management of micro-economy system (Fig. 1.1). On the other side, millions of people in extreme rural captured area are leading stressful life with hunger and poverty (Fig. 1.2). The hunger is on rise. The absolute number of undernourished people, i.e. facing chronic food deprivation, has increased to nearly 821 million in 2017, from around 804 million in 2016 (Table 1.1).

Currently, a lot of families across the globe struggle to feed their children a nutritious meal. In a world where we produce enough food to feed everyone, out of 821 million people, one in nine, still, misses the night meal every day. Even more surprising is the fact that one in three suffers from some form of malnutrition. So, it is necessary to resolve the problems of food security and poverty in rural area so that minimum social justice will be accessed to the rural people deprived of regular food and shelter. Hence, it is important how to develop sustainable foundation in extreme rural location with the help of indigenous resources to confront with problems like hunger, malnutrition (Fig. 1.3), chronic health problem by structural transformation of prevailing extremely poor social economic transition and awfully deprived lifestyle with scarcity of minimum food and shelter and bring them in sustainable form for better life for tomorrow.

Presently, millions of children on different geographical localities under extreme climatic condition are even deprived of night meal before going for sleep (Fig. 1.4a, b). The same with the poverty deprived children go to school without breakfast and try breathlessly for primary education. More than one in three suffers from some

B. K. Behera et al., *Move Towards Zero Hunger*,
https://doi.org/10.1007/978-981-32-9800-2_1

Fig. 1.1 Discarded food is the biggest single component of US landfill and incinerators, according to the Environmental Protection Agency. (Source: Photograph: Dan Tuffs/for the Guardian

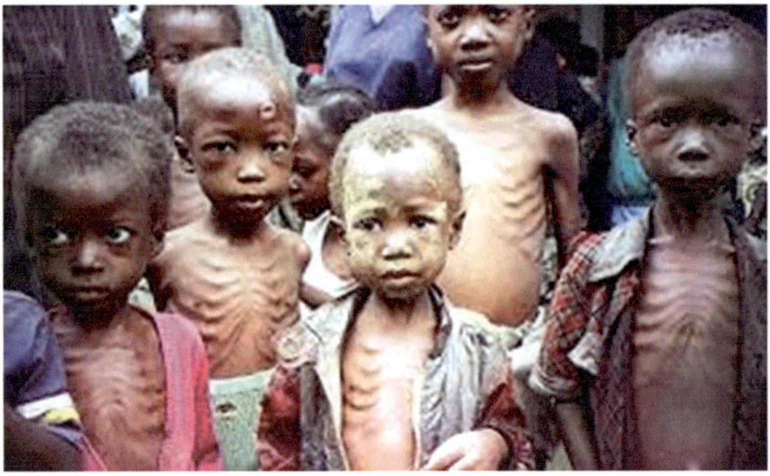

Fig. 1.2 Sources: http://www.usnews.com/news/blogs/data-mine/2014/12/31/un-wants-to-end-poverty-hunger-by-2030

form of malnutrition. For complete eradication of such problem, the work force struggling for zero hunger should properly understand the newly developed innovative methods/technology.

Table 1.1 Some important milestones in sequential development of food security majors

Food security tools utilised at specific reference periods	Nature of food security majors	References
Old Testament (1200 and 100 BC)	The idea of storing surplus grain in good times to guard against famine dates back at least as far as the Old Testament, when Joseph gave just such advice to the Pharaoh. Its history in China is almost as long, and ancient records describe how the emperor's "Ever Normal Granary" not only prevented famine but also allowed the state to stabilise prices for the benefit of both farmers and consumers	Old Testament
Sixth century	For hundreds of years, indigenous populations in the Philippines and Indonesia have been tapping the sweet, sugary sap of palm trees and feeding them to animals as their primary energy source	Camille Mendizabal
1896	USDA developed and implemented policies to control and proper utilisation of excess agricultural products by monitoring price, export support, developed food aid programme during post World War, 1918	Shaw [54]
1920–early 1930	Global price of agricultural products reached at minimum level and created havoc in export market. Act on Grain Stabilization Board was established for providing subsidies to agricultural products exporters. A Commodity Credit Corporation (CCC) was established to serve exporters unable to meet the global price	
1930	Accepted food security as global crisis. Globalisation of food. During World War I League of Nations started a joint venture programme on food security. Yugoslavia, as a member of the League of Nations, started perusing the problem of food insecurity at global level	Shaw [54]
1935	League of Nations submitted report on "Nutrition and Public Health" mainly concerning on food shortage and malnutrition. The multidisciplinary nature of food security was realised. The ultimate action of League of Nations resulted in developing modern food security tools and concept	Shaw Johan [54] and Boyd Orr [55]
1943	Historical Hot-Spring Conference was held at Virginia, USA, to form "Food and Agriculture Organisation" in order to increase food production and reduce the dependency on foreign resources, after post-World War II	Hot-Spring Conference

(continued)

Table 1.1 (continued)

Food security tools utilised at specific reference periods	Nature of food security majors	References
1945	FAO (Food and Agriculture Organization) made the first "World Food Survey" to find out whether there was enough food and macronutrients for everybody on the earth. It was noticed that one third of the world population was deprived of sufficient energy for the survival of (in 1945) life	The Food and Agriculture Organization of the United Nations [57]
1949	US government established "Section 416"an important source of funding to procure surplus commodities as food aids	Simon [58]
1952	Committee on Commodity Problem (CCP) was established to survey on how increase in agriculture production resulted in generation of surplus food commodities	Bhawmick [59]
1954	US government established "Public Law (PL) 480" to finance food aid activities	Simon [58]
1960	"The provision of Food Surpluses to Food-Deficit people through the United Nations Systems" (UN1961) by the US General Assembly to establish World Food Programme	Simon [58]
1961	World Health Organization (WHO) and FAO jointly created the joint Codex Alimentarius Commission in order to regulate the food safety by establishing international standards regarding processing, labelling, sampling of analysis, hygienic requirements, etc. of food commodities	Simon [58]
1963	General Agreement on Tariff and Trade (GATT) was established to understand the importance of the international trade of agriculture products	Simon [58]
1967	International Grains Agreement was approved at a conference called by the International Wheat Council and UNCTAD in Rome. This agreement was made of two different conventions: the International Wheat Agreement and the Food Aid Convention	Simon [58]
1969	The FAO Committee on Commodity Problems (CCP) approved the CSSD Catalogue of transactions which de facto was defining food aid	Simon [58]
1972–1974	During 1950–1960 the world food production was 50% higher than earlier production. However in 1972, due to bad climate condition resulted in dramatic increase in export value of cereals from United States	Simon [58]

(continued)

Table 1.1 (continued)

Food security tools utilised at specific reference periods	Nature of food security majors	References
1974	The United Nations World Food Conference took place in November 1974 in Rome, as one of its objective, to agree on measures to ensure that "within a decade nobody would suffer from food insecurity". The conference approved an International Undertaking on World Food Security which for the first time recognised that food security was a common concern of all nations	Simon [58]
1979–1980	Refugees from Afghanistan and Cambodia and forcing the international community to deal with protracted refugee, and hence emergency, situations and thus to reconsider in particular the modalities of their food aid operations	
1981	New concepts entered into the debate following research made on famines by the Nobel Prize winner, the Indian Amartya Sen, whose major argument was that during the past famines, the main problem was not so much the lack of food but rather the impossibility for poor people to access it	
1983	In 1983, the FAO Conference adopted a resolution on World Food Security (FAO 1983) which stated that "the ultimate objective of world food security should be to ensure that all people at all times have both physical and economic access to the basic food they need"	
1983–1985	Drought in Africa caused another challenge for the affected countries and the international community to deal with food security in case of shocks	
	The access dimension, however, as highlighted by Sen in 1981, was formally recognised in 1996 only at the Rome World Food Summit and practically put into practice by food security practitioners only after the 2005 Niger crisis and the 2008 world food prices crisis. Why did it take so long? Why were opportunities lost? A number of them occurred during this period	
1992	The 1992 International Conference on Nutrition, jointly organised by FAO and WHO, met in Rome and was a major milestone in the recent development of food security. The Conference approved an ambitious Plan of Action which called for inter-sectoral cooperation and coordination between all actors concerned. (FAO/WHO 1992)	

(continued)

Table 1.1 (continued)

Food security tools utilised at specific reference periods	Nature of food security majors	References
1996	After 1974 World Food Conference, again, World Food Summit was organised in 1996 by FAO in Rome	1."Declaration of the World Summit on Food Security, FAO Web site, 16 November 2009" (PDF). Retrieved 2010-10-15
	The Summit is and remains a major milestone in the history of food security	
1990–2005	Golden years of food security	
2002	The World Food Summit in Rome passed resolution to form an intergovernmental working group to prepare a set of guidelines on the implementation of the right to food. This resulted in the drafting of the Right to Food Guidelines	Background to the Voluntary Guidelines [60]
2009	The World Summit on food Security took place in Rome, Italy, between 16 and 18 November 2009. The decision to convene the summit was taken by the FAO Council, and unanimously adopted a declaration pledging renewed commitment to eradicate hunger from the earth at the earliest possible date	"Declaration of the World Summit on Food Security, FAO Web site, 16 November 2009" (PDF). Retrieved 2010-10-15

Shaw [54]
Boyd Orr [55]
"Declaration of the World Summit on Food Security, FAO Web site, 16 November 2009" (PDF). Retrieved 2010-10-15
The Hot Springs Conference [56]
The Food and Agriculture Organization of the United Nations [57]
Simon [58]
Bhawmick [59]
Background to the Voluntary Guidelines [60]

1.1.1 Genesis of Zero Hunger Concepts

In the opinion of Josue de Castro, in the year 1930, there was a lot of lacuna in the economic model of Brazil. The poor and economically deprived sectors were ignored while planning the economy of the country. There was a big gap lying between poor and rich. Out of 180,627,000 population of the country, 170 million were suffering under extreme food scarcity. Keeping in view of this social discrepancy, President Lula, in 2003, wanted to improve the handicapped social structure so that each and every individual should get minimum food and shelter to go ahead with a normal pattern of life. President Lula wanted to develop a new model of social structure and function on hunger eradication and social inclusion, linking macro-economic, social and productivity policies. His proposed model at the later

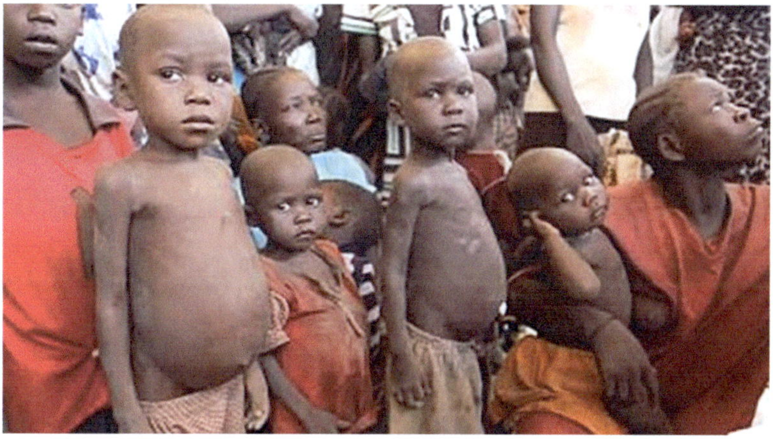

Fig. 1.3 Malnutrition in Niger reaches the emergency threshold. (Soure: UN)

phase was famous as "Zero Hunger Model". In this connection President Lula wanted to have favour from the entire nation to bring success in his great planning. Before him the great challenge was how to initiate the move for zero hunger, nation's food security and poverty reduction. For this purpose, from the beginning of his tenure, he started engaging all the sectoral ministries and level of government as well as the Brazilian society in general in a massive way: concentrated effort to work on zero hunger agenda. An integrated national food and nutrition security policy was framed under strict rule and regulation, and it was at later phase underpinned by a new legal and institutional framework.

The main motive of the government was to ensure that all Brazilians could enjoy their right to adequate food. Policies for raising production were linked to those promoting inclusion so as to enhance their effectiveness. So, like this way, the new buying power created by social protection was harnessed to stimulate increased food production by small-scale farmers who were poor themselves, thereby strengthening the local economies of their communities. In 2010, at the close of President Lula's second term, the *Bolsa Familia* (family Allowance programme) conditional cash transfer (CCT) programmer covered 12.8 million families, i.e. more than 51 million people. As of late 2010, Brazil's economy is the largest of Latin America and second largest in the America. From 2000 to2012, Brazil was one of the fastest-growing economies in the world.

Looking at the economic progress of Brazil, "Zero Hunger", the United Nations Secretary-General Ban Ki-moon launched campaign for "Zero Hunger Challenge" (ZHC) in 2012. The basic motto was "Achieving Zero Hunger is our shared commitment. Now is the time to work as partners and build a truly global movement to ensure the Right to Food for all and to build sustainable agriculture and food system" [1].

Fig. 1.4 (**a**) Crop land under extreme climatic condition. (Source: www.bread.org/what-causes-hunger). (**b**) Nigerians at risk with extreme poverty

The ZHC is mainly based on five elements:

(i) All food systems are sustainable: from production to consumption.
(ii) An end to rural poverty: double small-scale producer incomes and productivity.
(iii) Adapt all food systems to eliminate loss or waste of food.
(iv) Access adequate food and healthy diet, for all people, all year round.
(v) An end to malnutrition in all its forms.

Secretary General Ban Ki-moon wanted to integrate the five elements of the ZHC into nationally led SDGs which would be helpful in ending hunger and pay dividends right across a broad range of SDGs. In addition he was having intension to include each element in the 2030 Agenda.

As reported in the Global Hunger Index 2017, the world has made progress in reducing hunger since 2000; however, the progress has been uneven, and not fulfilling the claims, earlier. Recent report says that several countries have higher hunger level in 2017 than they did in 2008, indicating that progress in reducing hunger over recent decades is reversing more recently in some places (Fig. 1.5).

Due to scarcity of data bank and reference books, it has been a herculean task to resolve the problem of extreme poverty and go ahead with zero hunger in a substantial manner. So, it is highly necessary that there should be some ready references for social volunteers; NGOs; international organisations like United Nations Rome-based agencies (RBAs), the Food and Agriculture Organization of the United Nations (FAO), the International Fund for Agricultural Development (IFAD) and World Food Programme (WFP); policymakers; bureaucratic politicians; and students studying sociology, social sciences, agriculture, public health and veterinary to have inquisitive mind to know more about zero hunger and to find out a right direction to move for zero hunger, at a global level.

1.1.2 Strategic Plan for Food Security and Zero Hunger

The practice of food security is 10,000 years old, with central authority in civilisations ancient China and ancient Egypt being known as to release food from storage

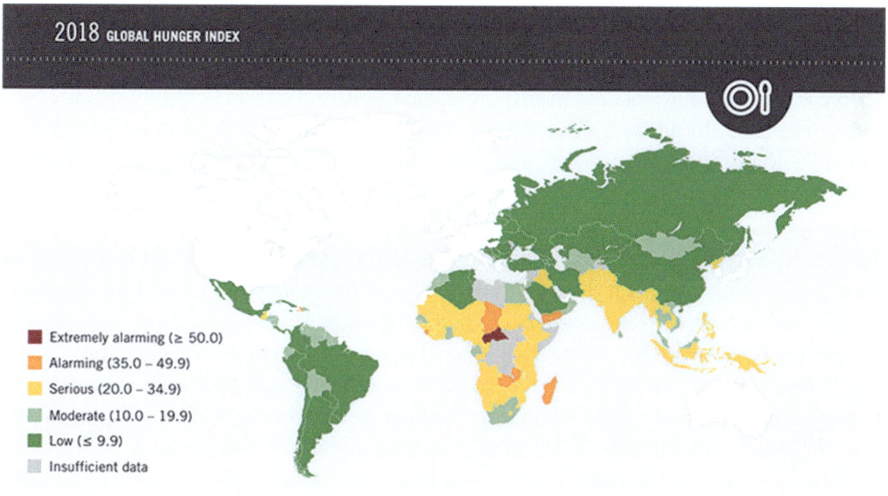

Fig. 1.5 2018 Global Hunger Index results – global, regional and national. (Source: https://www.globalhungerindex.org/results/)

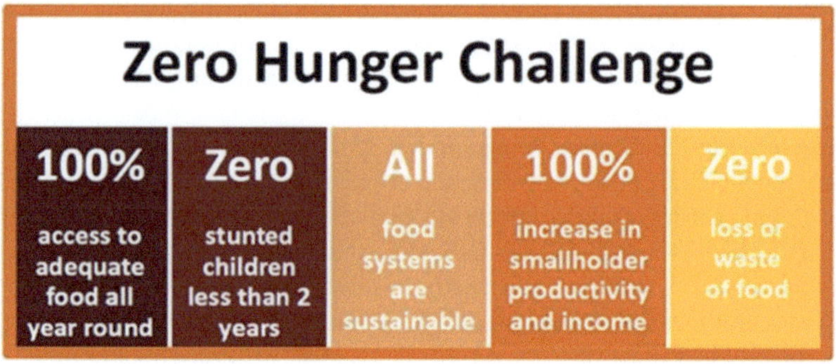

Fig. 1.6 Integrated Food Security Framework and its strategic plan of action for food security. (With courtesy from ASEAN, DHRRA News, Food Security)

in times of famine. In due course of time, and due to increased magnitude of population and demand of food, people started realising to develop some alternate majors for food storage which can be helpful during natural catastrophes such as floods, drought, earthquake and tsunami happening from time to time and causing damage to food, shelter and life. Since the last three decades, more attention has been paid to develop tools to reduce food insecurity (Table 1.1). In this regard contribution of institutes plays a major role in developing innovative projects for food aids with the collaboration of food security governance (Fig. 1.6).

The years 1990–2005 were famous as the golden years of food security, immediately after the fall of Berlin's wall in 1992 as the Anti-Fascist Protection Rampart (German: Antifaschistischer), when the Southern Africa was in the grip of tremendous drought with highly intensified food crisis. During this critical food crisis, at least 15 high-level international conferences were held to deal with food security and to recommend major multidisciplinary tools to fight against food insecurity.

The World Food Summit (WFS), 1996, described food security as: "When all people at all times have access to sufficient, safe, nutritious food to maintain a healthy and active life". In addition The WFS says food security is a functional practice in which "availability at all times of adequate, nourishing, diverse, balanced, and moderate world food supplies of basic foodstuffs to sustain a steady expansion of food consumption and to offset fluctuations in production and price" [2].

The WFS came into conclusion that suitable food aid majors should be framed to get rid of famines and starvation. It also decided that everyone should have access to safe and nutritious food, consistent with the right to adequate food and fundamental right of everyone to be free from hunger. It was also decided that "each nation must adopt a strategy consistent with its resources and capacities to achieve its individual goal and, at the same time, cooperate regionally and internationally in order to organize collective solution to global issues of food security".

Subsequently, in 2002 and 2009 two more food security conferences were held in connection with eradication of hunger at global level. The main target of these conferences was the physical and economic access to food which can fulfil the dietary as well as their food preference.

The World Food Programme (WFP) is an international voluntary organisation promoting food security and campaigning for zero hunger movement worldwide. It also intends to develop the quality of nutrition to improve the quality of life to the most vulnerable population and promote self-reliance. The WFP is in process to develop new infrastructure, or learn new skill, that will increase food security, in exchange for food. In this connection WFP has a partnership with the Government of Kenya in *Food Assistance for Assets* programme in order to bring food security for the nations suffering from inadequate food for daily life.

WFP assists in providing food and nutrients to an average of more than 90 million people in 80 countries, where extreme food shortage and malnutrition problems are prevailing [3–5]. It is a member of the United Nations Development Group.

The Food and Agricultural Organization (FAO) conference held in 1961 proposed to establish the World Food Programme (WFP) to ensure global food security. Finally, this international voluntary organisation was established in 1963, by the FAO, and the United Nations General Assembly on a 3-year experimental basis. In 1965 this programme was conformed on regular basis.

The main target of WFP is to move for zero hunger and bring solution to malnutrition being suffered due to poverty and extreme food scarcity. Besides this WFP is aimed to fight for micronutrient deficiencies, reduce child mortality, improve maternal health and combat diseases, including HIV/AIDS. Under the scheme of "food for work programmes", WFP has provided food to the people while working on projects like rebuilding damaged infrastructure and replanting crops following crisis.

The objectives that the WFP is expected to achieve are to [6]:

1. "save lives and protect livelihood in emergencies"
2. "Support food security and nutrition and (re)built livelihoods in fragile setting and following emergencies"
3. "Reduce risk and enable people, communities and countries to meet their own nutrition needs"
4. "Reduce under nutrition and break the intergenerational cycle of hunger"
5. "Zero Hunger in 2003"

Besides this, WFP intends to cooperate and help the other needy countries to fight for adequate availability of micronutrients, reduce child mortality, improve maternal health and combat disease. Food for work programmes help promote environmental and economic stability and agricultural production.

Generally WFP receives funds from various government sectors, international NGOs, corporations and private donors. WFP believes in partnership work in order to manage critical and emergency development of deficiency of food for the people of any geographical located place where unusual natural calamity brings food shortage problem. Many other international agencies such as the Food and Agriculture Organization (FAO), UN High Commissioner for Refugees (UNHCR), UK Department for International Development (DFID) a sister concern of UNHCR, EuropeAid and USAID; nongovernmental organisations such as ECHO (Educational Concerns for Hunger Organization), Hungry Kids organisation, Catholic Relief

Services, Save the Children and Norwegian Refugee Council; as well as corporate partners such as TNT, Citigroup and Boston Consulting Group are in process of serving food starved people suffering from unusual food scarcity and different malnutrition problems. In 2006, WFP developed partners with 84 different corporate and private entities and over 3000 nongovernmental organisations to serve the poverty captured area of global location.

Campaigning for eradicating food and nutrition deficiency in extreme poverty captured area of different geographic localities was also initiated by the World Health Organization (WHO) which was established on 7 April 1948, Geneva. It is a member of the United Nations Development Group. At present WHO also works to promote the availability of safe healthy and wholesome food for everyone in order to improve global food and nutrition security.

The WHO in collaboration with the World Bank has developed the International Health Partnership (IHP+) programme consisting of a group of partner governments, development agencies, civil society and others committed to improving the health of citizens of developing countries. WHO has also targeted to improve the quality of health and framed Alliance for Health Policy and Systems Research to improve health policy.

The Third International Conference on Financing for Development was held at Addis Ababa, Ethiopia (13–15 July 2015), and passed the resolution to help the poverty captured nations suffering from hunger from economic, social and environmental points of view by implementing the global sustainable development agenda at global level. Under the leadership of Ethiopian Foreign Minister Tedros Adhanom Ghebreyesus, the groundbreaking agreement, the *Addis Ababa Action Agenda*, passed resolution for enhancing global partnership to foster universal, inclusive economic prosperity and improve people's well-being while protecting the environment. In order to implement sustainable development project, the Addis Ababa Action Agenda contains more than 100 solid measures mainly covering technology, science, innovation, trade and capacity building. In addition, priority target is given to sustainable development of infrastructure, social protection and technology by financing energy, transport, water and sanitation and other areas to help realise the proposed sustainable development goals.

In 2015, the United Nations General Assembly passed resolution on Sustainable Development Goals (SDGs) or Global Goals for Sustainable Development having 17 global goals on economic development issues including poverty, hunger, health education, global warming, gender equality, water, sanitation, energy, urbanisation, environment and social justice. The SDGs came into effect in January 2016, and they will continue to guide UNDP policy and funding until 2030. The basic plan of SDGs is to focus on key issues including poverty alleviation, democratic governance and peace building, climate change and disaster risk and economic inequality. The UNDP helps the government to integrate the SDGs into their national development plans and policies.

For a zero hunger way of healthy life, and facing the challenges of wasting less food, eating better and adapting a sustainable lifestyle, the world is in a big dilemma how to ensure the food security of about ten billion population which will be reaching at the end of 2050. This uncertainty would have been due to the blue planet experiencing increasing water and land scarcity, soil, land and biodiversity degradation and more frequent and severe unfavourable weather. In order to overcome these problems, food security requires an integrated approach to find out suitable majors for solving malnutrition, resilience of food systems and proper management of biodiversity and genetic resources for sustainable agricultural productivity.

In 2018, three United Nations agencies, i.e. UN Food and Agriculture Organization (FAO), International Fund for Agricultural Development (IFAD) and World Food Programme (WFP), are tasked to eradicate hunger and ensure rural development by supporting food security and rural economic. These three international volunteer organisations have signed a 5-year memorandum of mutual understanding. The main reason of this memorandum is to work together with more collaboration, coordination and synergies, especially at country level to develop better support in serving the member countries to work out successfully in the implementation of the 2030 Agenda Sustainable Development. Although FAO, IFAD and WFP each have a unique mandate related to technical expertise, international financial assistance and emergency food assistance, the three Rome-based agencies share a common vision to achieve food and nutrition security and sustainable and sustainable agricultural development for ending hunger. The basic contribution by this mega collaboration is to bring commitment into action and action into impact and results. The agenda include, on priority basis, to increase aggregate investments which will be helpful in better productivity, employment opportunity and improvement in the activity of small-scale producers while offering broader opportunities for the poor and vulnerable. In addition priority is also given to reduce the intensity of poverty and hunger through social protection. Well-managed practice for social protection would be helpful for the needy people in improving lifestyle, strengthening food security and reducing undernutrition events. Social protection is a powerful tool for increasing capacities and productive potential of the poor. Thus, combining social protection and pro-poor investment will be helpful in overcoming the chronic problem of poverty and hunger.

1.1.3 Social Inclusion for Zero Hunger

The practice of social inclusion is to give equal opportunity to those people suffering from deprivation of social equality in having normal lifestyle access to resources, voice and respect for human right.

The basic intension of integrating social inclusion in zero hunger drive scheduled for 2030 Agenda is for complete eradication of poverty, hunger and undernutrition problem at global level (Table 1.2), and every person should reap the social advantage of prosperity and get a chance for minimum standard of daily lifestyle. Social inclusion is an unavoidable practice for the successful implementation of the 2030 Agenda Sustainable Development Goals targeted at freeing all nations and people and segment of society under the blanket of poverty, malnutrition and hunger. In addition, it ensures healthy lives and access to free education, adequate energy

Table 1.2 Prevalence of undernourishment in the world, 2005–2017

	Prevalence of undernourishment (%)					
	2005	2010	2012	2014	2016	2017[a]
World	14.5	11.8	11.3	10.7	10.8	10.9
Africa	21.2	19.1	18.6	18.3	19.7	20.4
Northern Africa	6.2	5.0	8.3	8.1	8.5	8.5
Sub-Saharan Africa	24.3	21.7	21.0	20.7	22.3	23.2
Eastern Africa	34.3	31.3	30.9	30.2	31.6	31.4
Middle Africa	32.4	27.8	26.0	24.2	25.7	26.1
Southern Africa	6.5	7.1	6.9	7.4	8.2	8.4
Western Africa	12.3	10.4	10.4	10.7	12.8	15.1
Asia	17.3	13.6	12.9	12.0	11.5	11.4
Central Asia	11.1	7.3	6.2	5.9	6.0	6.2
South-eastern Asia	18.1	12.3	10.6	9.7	9.9	9.8
Southern Asia	21.5	17.2	17.1	16.1	15.1	14.8
Western Asia	9.4	8.6	9.5	10.4	11.1	11.3
Central Asia and Southern Asia	21.1	16.8	16.7	15.7	14.7	14.5
Eastern Asia and South-eastern Asia	15.2	11.5	10.1	9.0	8.9	8.9
Western Asia and Northern Africa	8.0	7.1	8.9	9.3	9.9	10.0
Latin America and the Caribbean	9.1	6.8	6.4	6.2	6.1	6.1
Caribbean	23.3	19.8	19.3	18.5	17.1	16.5
Latin America	8.1	5.9	5.4	5.3	5.3	5.4
Central America	8.4	7.2	7.2	6.8	6.3	6.2
South America	7.9	5.3	4.7	4.7	4.9	5.0
Oceania	5.5	5.2	5.4	5.9	6.6	7.0
Northern America and Europe	<2.5	<2.5	<2.5	<2.5	<2.5	<2.5

[a]Projected values Source: FAO

availability and being free from social exclusion. The achievement of complete success in integrated institutional project plays a key role in order to have efficacy in promoting the rule of law, ensuring equal access to justice and broadly fostering inclusive and participatory decision-making. So, by realising the importance of social inclusion, President Lula included non-contributory social protection police while framing four programmes for Bolsa Familia. In 2004, the Bolsa Familia was integrated into zero hunger under the new Ministry of Social Development and Fight against Hunger. It was mainly designed to develop management of non-contributory social protection policies. Since the zero hunger (ZH) strategy was launched in 2003, the attention on Brazil's social protection policy has changed remarkably from initial emphasis on food nutritional security to economic growth with social inclusion. The social protection policy has two advantages: by increasing wage levels and stimulating human capital development in the poorer population.

The UN opined that proper attention should be paid for social inclusion and pro-poor investments in poverty captured area of different geographic locations to achieve zero hunger, globally. Social inclusion measures can be used to reduce the poverty gap in between earned incomes and the income necessary to procure basic needs for normal pattern of lifestyle. The agriculture and infrastructure development in pro-poor sectors would be immensely helpful in food security. In addition, it has been observed investigation that "if women farmer are given access to resources as men, the number of hungry could reduce by as many as 150 million". So, it is high time to realise and develop a gender-balanced world, where women have equal access to all types of social fronts like education, land, inheritance, resources and wages to support the zero drive movement.

In active social inclusion practice, each and every citizen has the right to participate in all sort of social activities including job opportunity that will lead to a normal life with food and shelter. Under this programme adequate income support with help to get job is given first priority. Inclusive labour market is also a part of active social inclusion in which people will easily access to join the workforce tackling in-work poverty, avoiding poverty traps and disincentives to work. The active social inclusion supports access to quality services, including getting back to work.

By practising active social inclusion, different nations want to tackle problems like poverty, social exclusion, in-work poverty, labour market segmentation, long-term unemployment and gender inequality. For example, millions of Europeans are under the umbrella of intense poverty due to worsening social situation caused due to economic crisis undermining the sustainability of social protection systems. In order to overcome poverty and develop effective social inclusion practice, a joint commitment among national governments, EU institutions and key stakeholders is taken into confident to fight against poverty and social exclusion with the target of achieving goals in 2020. The European government is in constant touch with EU stakeholders, NGOs, social partners, businesses and social economy actors, academics, foundation and international organisations to ensure their involvement in framing and implementing policy initiatives to combat poverty and social exclusion.

Effective implementation of social inclusion for empowering women will reduce the problems of hunger, poverty and malnutrition to a major extent. Rural women form the backbone of rural societies and play varied and vital roles as farmers, farm workers, entrepreneurs, caregivers and community leaders. Women are committed to food security to their families and their communities by their vital contribution to food production, food processing and marketing. It has been understood that 45% of women from developing countries take part as labour force. Women possess high potential in fostering sustainable agriculture diversification, promoting bio-fortification, reducing food loss and waste and helping in food processing for improved nutrition and food safety. So, in house improvement of women's social and economic condition will have direct impact on food security and nutrition, in particular on child nutrition.

Facts and figures from most of the developing countries like Africa, Asia and Latin America have clearly explained that the women's access to productive resources, agriculture technologies, market opportunities, social protection and education will be helpful in zero hunger movement for spending normal life. Practice of social inclusion for women and giving them opportunity in social decision-making within households, communities and rural organisation and even at government sector will be the most effective and favourable factors for eradication of hunger, malnutrition and poverty.

1.1.4 Resilience-Building Approach

With the passing of time the habitable world has been facing variety of risks and unpredictable happenings which challenge the human community for better survival with adequate food and proper shelter to live, in a sustainable pattern (Fig. 1.7). Mainly, natural calamity, community disease, variable and volatile prices and political instability are some of the major adverse factors responsible for leading hurdle life which often lead to hunger and poverty. In 2017, almost 124 million people across 51 countries and territories faced crisis levels of acute food insecurity and were in need of immediate humanitarian action [7, 8]. Currently at global level, about 52 million children were acutely under extreme malnutrition due to cyclical food insecurity and protracted crisis exacerbating vulnerability. This is far off the internationally agreed global nutrition target to reduce and maintain childhood wasting below 5% by 2025 [8]. The 2017 ruthless food crisis in South Sudan, Nigeria, Somalia and Yemen was an extreme chaos among international organisations to find out immediate solution and save hundreds of inhabitants to get relief from this catastrophic situation. But unfortunately humanitarian requirements remained unmet, and longer-term investments were far below projected needs.

Philanthropic attitude towards social crisis due to unusual mishappenings, and threat survive by creating extreme food scarcity, poverty and malnutrition like chronic social problems. A resilience-building approach will be immensely helpful in mitigating the destructive effects of shocks and stressors before, during and after crises, thus reducing misery prevailing in the society [9, 10]. The United Nations Office for Disaster Risk Reduction (UNISDR) defines resilience as: "The ability of a system, community or society exposed to hazard in a timely and efficient manner,

Fig. 1.7 Sustainable Development Goal 2: End hunger, achieve food security and improved nutrition and promote sustainable agriculture. (Source: https://unric.org/nl/sdg-2)

including through the preservation and restoration of its essential basic structures and function" [11–13].

After realising the importance of resilience and its role in reducing risk in poverty and hunger captured nations, the Food and Agriculture Organization of the United Nations (FAO), International Fund for Agricultural Development (IFAD) and the World Food Programme have developed a common approach to building resilience to eradicate hunger, poverty and malnutrition. These three organisations are collectively known as Rome-based agencies, and their main goals include agriculture, food security and nutrition; resilience is essentially about the efficacy of individuals, groups, communities and institutions to discharge the responsibility of food aids timely with adequate quantity to face shocks related to food insecurity or at risk of becoming so.

Resilient livelihoods are the most essential for the people who are susceptible to hunger and poverty. Sometimes, unpredictable climatic condition as happening in Zimbabwe with extreme scarcity of seasonal rainfall created food insecurity and undernutrition problem leading to hunger and poverty. In order to overcome such crisis, WFP's 2016–2020 for Zimbabwe has taken major steps for food security and nutrition challenges having focus on long-term recovery and resilience building to relieve from food insecurity while maintaining strong capacity in humanitarian assistance. This action taken is a practical example of a move for zero hunger and which may be applicable in the future as a model system to have control over unpredictable food crisis.

WFP's 2016–2020 strategy for Zimbabwe addresses food security and nutrition challenges, with a focus on long-term recovery and resilience building to remedy the underlying causes of food insecurity and undernutrition while maintaining strong capacity in humanitarian assistance. Rome-based agencies (RBAs) have also identified some significant social welfare majors which can be helpful in their venture in resilience building. Local and national ownership and leadership to ensure action plan strategies on their capability in resilience building to have control over hunger; multi-stakeholder approach for strengthening of vulnerable people's resilience; combining humanitarian relief and development during crisis and shock for immediate relief programme and also long-term majors for building sustainable resilience for food security; and mainstreaming risk-sensitive approaches for highly effective decision-making at government sector and international NGOs and strengthening them in coordination for building resilience in sustainable pattern are some of the important steps for zero hunger movement. The main targets for resilience building either before or after some natural disaster are to develop pillaring strength in building resilience which could be helpful in absorptive, adaptive and transformative capability of hunger captured populations. These responses could bring sustainable relief in tolerating shocks and social hurdles in a stressful situation by developing potential for adaptability to new social amendments and which will be helpful in transformative potential to transform the set of livelihood choices available through empowerment and growth. The transformative framework on infrastructure, community network, ability for empowerment and development, social inclusion and social protection will allow for a habitable extreme geographic located nation or a part of nation to give relief from shocks and constraints.

1.1.4.1 Government and Governance

It is obvious that health, accesses to basic education, economic instability, climatic changes and threat to environment and ecosystem are not directly under our control; however to respond to these challenges, how we absorb the unpredictable shocks and how quickly we recover after a blow depend on our collective response for building resilience. Building resilience for threats and risk to food insecurity and nutrition deficiency depends on the nature of shocks generated by unpredictable natural adverse climate and calamity. So, it is necessary for decision-makers in government to develop resilience-based policies, legal frameworks and implementation programmes mainly targeting (i) providing policy advice, (ii) building local potential and strengthening nation coordination and (iii) involving strategic actors from the private sector and civil society organisation. One of the best examples of working capability of government and governance is the formation of FANRPAN (Food, Agriculture and Natural Resources Policy Analysis) which was established in 1994, when the Ministers of Agriculture from Eastern and Southern Africa realised the need for an independent policy input to address chronic food insecurity and the challenges of managing natural resources. In 2003 FANRPAN was officially registered as a private organisation (PVO) in Zimbabwe. Its main function is sharing knowledge and acquisition of agricultural and food security policy analysis and advocacy, as well as research and formulation of priority Africa agricultural research, initially in Southern and Eastern Africa, and latterly extending its contribution and reach to continental and global levels. Presently, FANRPAN is present in 17 African countries with the network that includes universities, research institutes, the business sector, farmer groups and other civil society organisations that have a stake in FANR policies. In 2016 FARPAN has been ranked 13 out of 92 in sub-Saharan Africa and 55 out of 17 globally in the 2015 Global Go to Think Tank index Report (TTI), led by the University of Pennsylvania through its Think Tanks and Civil Societies Program (TTCSP).

The nation's food security depends on governments who can play primary actors in the physical, social and economic aspects. Any efforts in increasing agricultural development and food security outcomes must be considered the role of governance. This will be helpful in establishing more stable and transparent governments which will ultimately be effective to agriculture and food security systems. In order to reduce chronic hunger the intricate connection between agriculture, food security and governance should be cut down, and a straightforward relationship should be developed to increase the efficacy in prompt action by giving freedom in action to seize food insecurity and social exclusion.

Integration of resilience building with governance has been gaining confidence rather than traditional approaches to solve global food crises. Perfect coordination among government, governance and societies has found support in instances where food security work coincides with other efforts to improve governance. Food crisis between 2007 and 2008 virtually was a big threat to governments as well as social stability in Africa, Asia, the Middle East and Latin America and Caribbean. From this it was realised that food security required good governance at international, national and local levels. The only way left to overcome such situation is effective coordination of governance, food security and agriculture work at global level.

1.1.4.2 Governance and Agriculture

Currently, about 12% of the global population is under chronic undernourishment. In addition, one third of food aids have been in waste every year. A wide social exclusion is prevailing in the people who live with chronic hunger and extreme poverty [14–16].

A lack of food security results in chronic undernourishment for over 12% of the global population. At the same time, a third of the world's food supply goes to waste every year. To make matters worse, people who live with chronic hunger and extreme poverty are often excluded from political representation, government services and government benefits.

At present the policies implementation and practice is lacking of coordination between government, private sectors and farmers, as agriculture and food security are hindered by complex political process. This is mainly due to the poor governance responsible for food insecurity management. Historically speaking most of the armed conflicts in the past were due to low income of food-deficit countries which depend on domestic agriculture production. Agricultural practices are often harmed by conflict and improper framing of government policies.

1.1.4.3 Governance and Food Security

Food security and population are complementary factor for successful governance having effective rule of law. Populations under the umbrella of food security have confidence in participating in political dynamic processes. On the other hand food security and poverty are mainly responsible for grievances raised by public against institutions, hinder political participation and contribute to social unrest. The only way out left for government's legitimacy and stability is by removing social-political obstacles and enhancing food security. A food-secure population can bolster stable governance, whereas a food-insecure population can destabilise governance.

1.1.4.4 Entry Points for Food Security Initiatives

To have control over food insecurity and make the world free from hunger, integrated multidisciplinary initiatives such as Feed the Future 2030 and the Milan Food Policy Pact can play well the role in solving food crisis in an integrated way. This type of organisation can have multipurpose resolution and potential to act for inclusive rural transformations in order to establish balance link between rural-urban growth and development. Currently, involvement of governance for integrated agricultural development is catching up attention among farmers. International agencies like FAO, the Organization for Economic Cooperation and development (OECD) and the United Nations Capital Development Fund (UNCDF) have taken initiative to resolve issues related to food insecurity and eradicate hunger, at global level. These organisations are well experienced in analysing the territorial policies and governance system for food security in Cambodia, Colombia, Cote d'Ivoire, Mali, Morocco, Nigeria and Peru with the support of their governance. The process of social inclusion and decentralisation of governance can accelerate in solving local food security issues and give opportunity to women and vulnerable populations having small-scale agricultural activities. Such practices peruse the policy-makers, and private sectors having all sort of powers and resources have to be accountable before the people at all level. For example, through a USAID funded project, citizens in Yene, Senegal, were involved in a party stakeholder dialogue on governance in land management which ultimately helped to improve the management of arable land and encourage the farmers in performing better agriculture practice. Thus, involvement of governance with agriculture and other social inclusion measures effectively helps in eradication of hunger and poverty of developing countries, at global level.

1.1.4.5 Food Security Information System

In order to have control over building resilience, facts and figures resulted from food insecurity shocks and threats are pre- and post-analysed. Proper analyses of data related to food crisis help in facilitating action process and accelerate building resilience, timely.

1.1.4.5.1 Resilience Social Habitat Approach

The international welfare agencies (NGOs) should provide cooperation in strengthening the capacities of governments and communities by providing immediate and efficient assistance for adequate supply of food aids in response to agricultural threats and emergencies to save lives, promoting recovery by reducing the gap between food assistance dependency and self-reliance. Proper presentations in pre- and post-shocks in food crisis situation and developing cooperation with food aids providing private and government agency help in accelerating recovery from risks for social instability from economic condition.

1.1.5 What Causes Hunger?

Although the mishappening of hunger is in a process of declining at global level, but, still about 795 million people are under the cloud of hunger and struggling for life for minimum food and food aids. At global level the food availability is enough to feed seven billion people, but it is surprising to learn that one person in eight on the planet goes to bed hungry each night. So why does hunger exist?

Many factors like natural calamity, unwanted and unpredictable climatic changes, war and conflict, extreme poverty, social exclusion, etc. are primarily responsible for the state of hunger, around the world (Fig. 1.8). Mainly, people from developing nations are hunger victimised, even from generation to generation in a cyclic order. The United Nations Food and Agriculture Organization (FAO) estimates that about 815 million people of the 7.6 billion people in the world, or 10.7%, were suffering from chronic undernourishment in 2016. Almost all the hungry people live in lower-middle-income countries. Children are the main victims of undernutrition which includes stunting, wasting, deficiencies of vitamins A and zinc; fetal growth restriction is a cause of 3.1 million child deaths annually or 45% of all child deaths in 2011(UNICEF, World Health Organization (WHO) and The World Bank 2018).

Goal two of the Sustainable Development Goals, developed in 2016, targeted for "Zero Hunger" by 2030. But still many developing countries like Africa are having

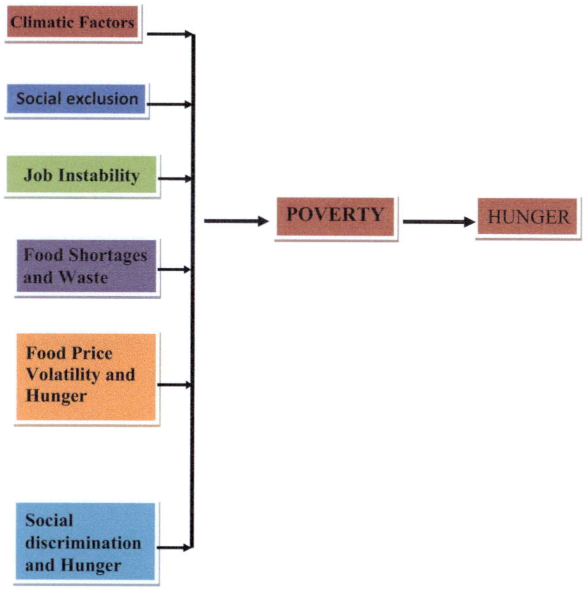

Fig. 1.8 Factors affecting poverty and hunger

hunger, as a main national problem. About 27.4% of the population in Africa was categorised as severely food insecure in 2016, which is almost four times as high as any other region. As reported by FAO 2017 the food security has been increased by about 3%. Alarmingly, the food insecurity is on a rising trend (Table 1.3). There are large differences among continents in the prevalence of severe food insecurity. Approximately 27.4% of the population in Africa was classified as severely food insecure in 2016, which is almost four times as high as any other region. Alarmingly, food insecurity is on the rise, specifically in sub-Saharan Africa. From 2014 to 2016, food insecurity increased by about 3% (Fig. 1.9).

1.1.5.1 Poverty

Poverty is one of the main causes for hunger, but it does not mean that poor people are in hunger; however, most of the hunger captured people are poor. Due to inability of affording for food and nutrition, million lives have been under the cloud of hunger with many suffering from malnourishment. The largest groups of the people in the world in extreme poverty are smallholder farmers in developing countries. Due to the lack of personally cultivated land, it is a dream for them to produce agricultural products for daily use.

Poor agriculture infrastructure and unfavourable trade laws are also responsible for hunger. Efficacy of governance and policies implementation boost agriculture production. For example, during 1930–2006, Vietnam was able to reduce poverty from 58% to 18% and became the second largest rice exporter in the world [3]. This could only be possible due to agricultural land reform and public approval of smallholder's agriculture in which crops were mainly grown on small farm instead of commercialised farming. Fair way of land distribution and investment in

Table 1.3 Percentage and number of people affected by severe food insecurity in 2016

	Percentage	Millions
World	9.3 (±0.4)	688.5 (±27.6)
Africa	27.4 (±0.7)	333.2 (±8.6)
Asia	7.0 (±0.6)	309.9 (±26)
Latin America	6.4 (±0.3)	38.3 (±2.0)
Northern America and Europe	1.2 (±0.1)	13.0 (±1.3)

Source: FAO, IFAD, UNICEF, WFP, and WHO. 2017. The State of Food Security and Nutrition in the World 2017. Building resilience for peace and food security. Rome, FAO

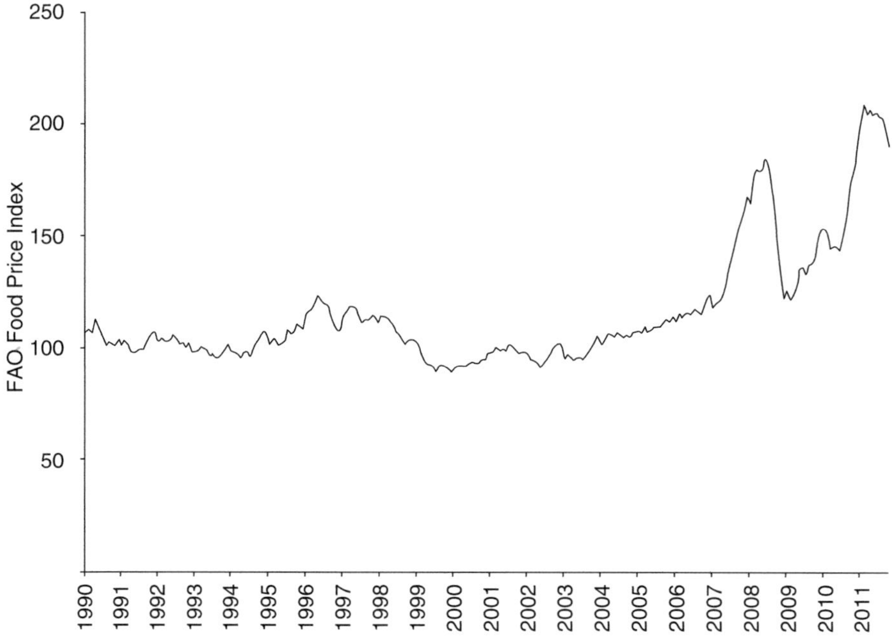

Fig. 1.9 World Food Price Index. (Food and Agriculture Organization annual real food price index 2002–2004 = 100)

agricultural infrastructure allowed Vietnam to increase food security and eradicate hunger to a greater extent [17].

People under poverty are lacking of nutritious food which makes them weaker, physically and mentally. Hunger and malnutrition are chronic problem in cyclic way, lasting for a long period. Children, who are chronically malnourished, or "stunted", often grow up to be adults whose incomes are lower. They are condemned to a life of poverty and hunger. In developing countries, also farmers are often not in a position to purchase seeds and fertiliser and unable to cultivate crop. In short, the poor are hungry and hunger traps them in poverty.

1.1.5.2 Job Instability

Even in developed country like the United States, an increasingly high number of children are suffering from hunger each day. Mainly, this is due to the parents of those children being jobless or having a job that doesn't pay well. This is also common in developing countries. Recently, the Department of Agriculture pointed out those 506,000 household children faced "very low food security".

Even in higher-income society, hunger is still prevailing due to job instability. Hunger rates rise when the national or local economy is in slump. Once the economy improves, some people continue to struggle to find work. For example, people who have been in prison face wide-scale discrimination that makes it difficult to find any work.

1.1.5.3 Food Shortages and Waste

One third of the global food production is left unused as waste when one in eight is hungry. Producing this much of food is equivalent to 3.3 billion tonnes of greenhouse gases, to the atmosphere, with consequences for the climate and, ultimately, for food production. Still, food shortages in developing countries are common. The people most affected are smallholder farmers and their families who depend on their own surplus to survive between harvests. The period leading up to a harvest is known as the "hungry season". Food from the previous harvest runs out and families cut back on meals. This period of time may last for months depending on the size of the previous harvest. Similarly, in the United States, families with very low incomes run out of money at the end of the month. Families cut back on how much they eat and then eventually skip meals altogether on some days. Another reason for food shortages is up to 40% of food grown in some countries is spoilage. Smallholder farmers do not have adequate storage facilities to protect their supplies against pests and weather.

1.1.5.4 Food Price Volatility and Hunger

Limiting food price volatility is the soul target of Sustainable Development Agendas (2030). Thus, it is utmost necessary to understand the causes and effects of food price volatility at global level. Still, poor information is available regarding the link between global food price volatility and local food insecurity. Besides this, it is also necessary to understand how different types of trade policy measures, which are a frequent political response to food shortages, affect global food price.

Mainly, food supply and distribution are controlled at global level, which ultimately influences regional markets. During 1961 import of cereals in global markets averaged at 26 kg per person per year, and by2009 it was increased to 46 kg per year per person. It was about 77% increases over five decades [18, 19]. In most of the developing countries, nearly half of the rural communities are smallholder farmers having a maximum of 3 ha cropland [3, 21]. Mostly, people of developing countries rely on retail agricultural market. This is about 58.7% in India, 60.6% in Indonesia, 74.8 in Kenya and 81.6 in Rwanda [21], vulnerable to price increase. These facts show that food price movements in global and regional markets have a significant impact on hunger. It has been estimated that the cases of undernourished people increased from 848 million to 923 million, mainly because of the raise in food price [22, 23]. Since the last decade, the international organizations have started showing awareness on the urgency of adverse affect of volatility of food price, and have also gone for proper action plan to resolve the issue on priority basis [24–26]. The World Bank, the Organization for Economic Cooperation and Development, the Food and Agriculture Organization, and other branches of the United Nations are in touch with update information on computerized data analysis and policy statements related to food price volatility [27]. Volatile prices challenge the governance and policy makers to manage food crisis, as and when arise (Fig. 1.10). The volatility of price vary according to change in food prices over time [27].

The Organization for Economic Cooperation and Development (OECD) and FAO say that in 2020 the world wheat and rice price will increase about 14–35% [20]. It has been predicted that future decades will witness more price spikes and the alleviation of hunger will depend on the responsiveness and vigilance of world governments [22]. The world population is going to be about seven billion at the end of 2050 [23]. The Asia-specific demand for more protein-rich food will require greater land and energy inputs. Meat consumption in developing countries has been growing at over 5% per year over the last few decades with an expectation of continued further growth [28]. Due to the process of land degradation, about 12 million hectares of fertile agriculture land having the capacity of producing 20 million tonnes of grain are lost [25]. Further food production is going to be effected badly by climatic changes [29] (Fig. 1.10). Other commodity markets are also having unstable prices. The rising trend in market prices of energy and mineral commodity are similar as in

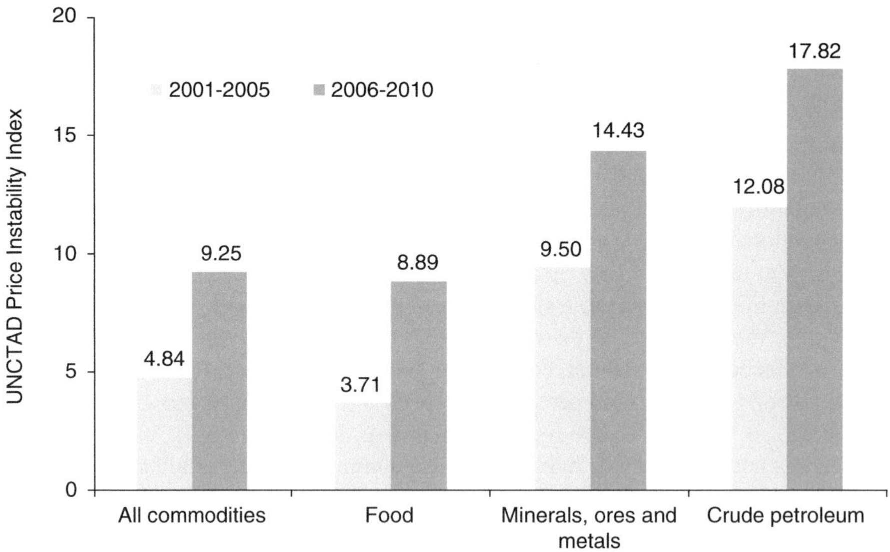

Fig. 1.10 Global commodity price instability index

the case of volatile market of food and food aids [26, 27]. The United Nations Conference on Trade and Development's (UNCTAD) global commodity price instability index (Fig. 1.10) shows the latter half of the previous decade (2006–2010) experienced greater volatility than the first half (2001–2005). The instability index is measured as the percentage of deviation of commodity prices from their exponential trend levels for a given time period.

1.1.5.5 Social Discrimination and Hunger

Every woman, man and children has the right to have sufficient food for survival and shelter to live. But a strong social exclusion is existing where female genders are excluded from social equality, mainly for adequate food and nutrition, to lead a normal life as compared to the male. In most of the developing countries, about 43% of women are agriculture labour without proper justification of payment as compared to male labour as counterpart. Women have human rights to be associated with each and every social activity, part of decision-maker and availing all sort of facilities related to food security. In most of the developing countries, women are involved in a variety of agriculture practices such as crops cultivation, livestock and fish farming and marketing. They produce food and cash crops at subsistence and commercial levels. At commercial level women are actively involved in all types of activities that support natural system management and agricultural development, such as soil water conservation, afforestation and crop domestication. Poor households headed by women often succeed in providing more nutritional food for their children than those headed by men. Social cultural traditions are more oriented for men's participation rather than women involvement.

Mothers having ill health and poor nutritional availability often give birth to underweight baby. The extensive differences in women's nutrition among developing countries having similar income levels indicate that something other than income affects women's nutrition. Inheritance cultural norms and practices on dietary habits often lead to poor health condition of women. For example, India and Bolivia are having similar gross incomes, but about 36% of Indian women are underweight, compared to 1% of Bolivian women [27].

Women confront a variety of hurdles to access productive inputs, assets to land and services necessary for rural livelihoods. These include access to fertilisers, livestock, improved seed varieties, extensive services, agricultural education and credit. Inequalities in ownership of access for social structure development affect women's food production and food security. Differences in risk and vulnerability between men and women cause food insecurity. The role of women in food utilisation is supposed to be the most critical from food conservation and food security point of views. Women are the expert in house management by social inheritance practices, and thus well experienced in maintaining the dietary quality by value added nutrition and food supplements. Access to food within the household is determined by cultural practices and power relationships within the family.

Women's role in food utilisation for food security is perhaps the most critical and outweighs the importance of their role in food production and how they spend the income they earn. Women are typically responsible for food preparation and thus are crucial to the dietary diversity of their households. Access to food within the household is determined by cultural practices and power relationships within the family. Although food may be available, adequate amounts to maintain nutritional intake may not necessarily be as accessible to women compared to men [27].

Food security and agricultural programmes can strengthen human rights and be more effective if they also tackle the constraints around women's access to resources. Human right includes right to education, knowledge and employment and women's equal right to land, water and other resources that are necessary for raising the productivity of their agricultural activity, food and nutrition security.

Irrespective of cast, religion, colour, ethnicity and demographic background, the food insecurity is problematic to each and everybody. Some groups, however, are more vulnerable than others. USDA reports that 22.5% of African American households and 18.5% of Hispanic households are food insecure, although both nations are higher than national average of 12.3% [30].

Despite years of progress, food security is still under threat. Policy implementation at governance level, discrimination in education and social exclusion are few important factors that cause hunger in minority population. Most of the people from developing countries are hunger victimised because of insufficient and poor pay structure which is not adequate to meet the monthly food budget, after paying money for rent and other utilities for daily requirement. Despite the gains that have been made since the civil rights movement, the unpleasant truth is that minorities (both men and women labour) still make up the majority of the low-wage workforce. The gap between wages of black, Hispanic and white workers continues to expand [31].

The wealth gap plays a huge role in creating food insecurity. Wealth is unequally distributed by race. The average wealth of white families is seven times higher than that of black families. It is because the minorities do not have much property due to the lack of sufficient wages and pay structure. In 2016, the family pension of an average white family in the United States was about $157,884 as compared to the family pension of Hispanic and black families amounting $53, 793. This is due to the reason that the lower-income group are under the constraint of poor pay structure, and are not able to save money for future security.

School education systems and the quality of education are directly linked with hunger. The UNESCO reports that "education provides skills that boost employment opportunities and incomes while helping to protect people from socio-economic vulnerabilities". The Alliance to End Hunger mentioned that "Schools with 90% white students spend $733 more per student than schools with 90% or more students of color". This sort of social discrimination in all sort of government facilities is prevailing in most of the developing nations like India.

The food security is also primarily caused by racial equity issue. A cloud of social racial discrimination is not only responsible for a healthy social structure but also responsible for food insecurity and poverty.

1.1.6 Pillars of Food Security

The sustainability in food security is prevailing in a society when the inhabitants have full access to social, physical and economic resources for adequate food and nutrition for an active and healthy life. Mainly food availability, access to sufficient nutritious food, utilisation and stability are the four main pillars of success of food security [32].

The quality and sustainability of food security is monitored when secure access to food synchronises with a clean environment, proper and efficient health services and adequate monitoring of healthcare services for active life for all household members [32, 33].

Different factors like agricultural production, existing infrastructure, international policies and gender and racial issues play a crucial role for sustainable availability of food in quality and quantity wise (Fig. 1.11). Proper quality educations are primarily responsible for bringing changes and improve food availability and access to food in a sustainable pattern. Food intake and health status are closely related to each other, and both determine the level and overall goal of food and nutrition security.

Mainly four factors for nutrition security are (i) availability, (ii) access, (iii) use and utilisation and (iv) stability which are responsible to frame the plan for food security.

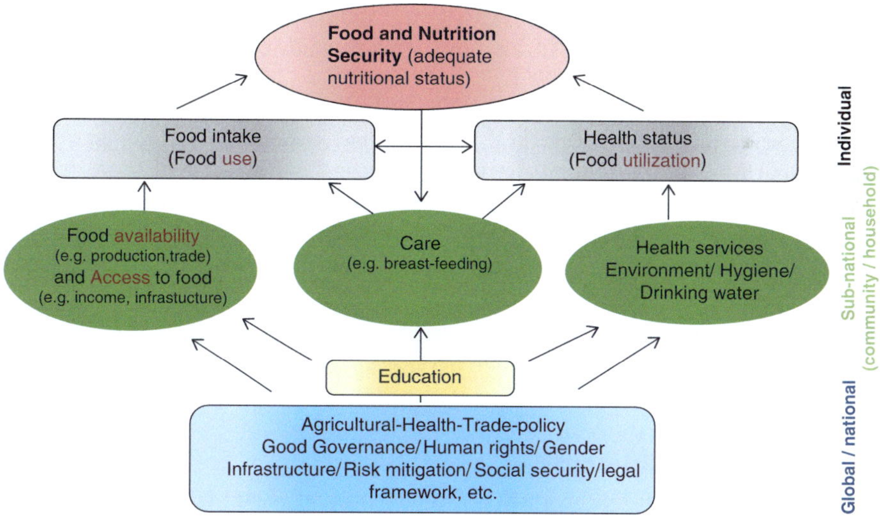

Fig. 1.11 The graph displays below shows the complex aspects and interplay of food and nutrition security on different levels. (Source: modified after UNICEF 1998)

1.1.6.1 Availability

The availability is defined as physical existence of food. On national level food availability represents domestic food production, commercial food imports and export, food aid and domestic food stock. Natural calamity and unpredictable climatic changes exert pressure on natural ecosystem, land and water resources. The impact of such phenomenon may cause land degradation, lack of irrigation water, reduce soil moister which bring imbalance in economic livelihood. Such changes in natural resources bring conflicts over usage of water resources, which are ultimately responsible for long-term food security. The Stockholm International Water Institute (SIWI) highlights about conservation majors and significance of green water (the water hidden in the ground as soil moisture) and blue water (water available in lakes, rivers and aquifers) [34]. In addition proper management of irrigation systems, up gradation of water-use technology, recycling of waste water, modernisation of soil-water management technology in rain fed systems and the resilience of agriculture systems can be strengthen and reduce risk in support of overall livelihoods [35, 36].

1.1.6.2 Access

Food access referred to the affordability and allocation of proper food to single family member or households for better living with sound health. The households should have sufficient access to diversified resources to possess food security both in quality and quantity wise. This mainly depends on the efficacy in functioning and implementing food security policies by governance. In the opinion of UN Committee on Economic, Social, and Cultural Rights (ESCR), the main causes of hunger and malnutrition may not be due to the scarcity of food but inability to access available food, usually due to poverty. Accessibility also depends on quantity of household resources and affordable price [36].

Access to food is mainly twofold, i.e. direct access and economic access. In former case household produces food manually, and economic access is mainly purchased food from foreign sources. A household access to enough food may not be able to fulfil the requirement of household members due to intrahousehold food allocation practice based on social inheritance cultural practice. Accessibility also depends on physical, social and environmental policy, and any abnormal changes in these factors may hamper the agricultural productivity and responsible for hunger. To save the life from such disaster and unusual happenings, construction of small dams and water reservoirs, rainwater harvesting systems and modernised water drainage systems to prevent flood are helpful [36].

1.1.6.3 Use and Utilisation

Use means the socio-economic status of household food and nutrition security as per the experience and habits. On the basis of quality and quantity of food availability, the households have to plan food procurement capacity and consumption to maintain good health without the problem like malnutrition. Proper biological utilisation of food mainly depends on the requirement of food and nutrition on the basis of physical labour of household members. This also required awareness of proper healthcare, nutrition value and proper storage system for timely availability of food. Clean drinking water is an unavoidable factor to prepare food, and directly intake of water, mainly at the time of any natural disaster or calamity. In most of the developing countries, the underground drinking water resource is contaminated with industrial effluents through soil bleaching process and causes alarming health conditions. About 884 million people worldwide have lack of any accessibility of filtered drinking water [31, 32].

1.1.6.4 Stability

On the basis of World Food Summit, the food stability means availability, access and utilisation of food at all time for all people, in a sustainable pattern. Chronic food security is disruption of the availability of food in cyclic manner irrespective of even any food crisis.

This mainly depends on the implementation of effective food security policy and food aids distribution among the needy people being managed by governance on various aspects like food aids distribution, income and economic resources availability and external risks such as natural disaster and climate changes, price volatility and conflict, improving the resilience of households. Besides the government should frame effective and efficient insurance policies to meet the emergency requirement during drought and crop failure, protection of environment and sustainable use of natural resources like land, soil and water [33, 34].

1.1.7 Challenges to Food Security

Challenges for food security are a universal phenomenon existing from insects to top predictor like human being. Each and everyone animate life of the habitable world is conscious over food security to overcome hunger in adverse climatic conditions. This practice has been well noticed in small insects even (Fig. 1.12).

Fig. 1.12 Insects are carrying food (as highlighted by arrow) for storing at the beginning of rainy season. (Photograph taken in Central Institute for Research on Goats, Makhdoom)

1.1.7.1 Water Scarcity

Due to scarcity of water and non-availability of irrigation canals, the farmers are under the constraint of pumping underground water with heavy diesel and electric pumps for irrigation purpose. This practice is also seen in large countries like China and India. This will eventually lead to water scarcity and cut back in grain harvest. In smaller countries like Afghanistan, Algeria, Egypt, Iran, Mexico and Pakistan also, the water scarcity problem badly affects the irrigation system, mostly under adverse climatic condition. Sub-Saharan Africa has the largest number of water-stressed countries at global level. Out of 800 million people who live in Africa, 300 million people are victimised by water scarcity problem [36]. It has been predicted that by 2030, 75–250 million people in Africa will have water scarcity problem [36]. In 1990 the World Bank has the global record of reclaimed Ica Valley desert in Peru into the largest supplier of asparagus in the world. Although asparagus cultivation provides jobs for local inhabitants, money flows to the buyers, mainly from the British. The main drawback of this project was creation of water shortage problem in the mountain region of Huancavelica.

1.1.7.2 Land Degradation

Land degradation is a global environmental problem threat to the habitable world on its greenery and agricultural productivity [31, 36, 37]. Globally, the land degraded area covers about 18.1 million km meters in which 92% was due to mismanagement and 38% is due to overgrazing. The land degradation has been a chronic practice caused by human activity for urbanisation and industrialisation. The importance of land degradation has been realised because of its direct link to food productivity leading to food insecurity. Degradation of land is not mainly due to population intensity but as the result of decline in soil fertility, soil structure, degradation of irrigated land and soil erosion. The severe land degradation, as noticed in recent past alarming the possibility of implementation of any sustainable development project, negative impact on socio-economic conditions, regeneration capacity of ecosystem and its profound impact to the wide spread poverty. Land degradation is mainly due to unsynchronised process of land quality and land use [38]. Land degradation may be a short-term process or long-term process, but responsible for reducing the productivity potential which ultimately causes deterioration of the natural way of conservation process of terrestrial ecosystems. In addition, land degradation is responsible for loosing biological diversity, economic productivity and threat to the longevity of crop land, pasture and woodland and forest ecosystem [39]. Globally, 20% of cultivated land, 30% of forest ecosystem and 10% of pasteurising land are under the grip of divesting process, which may ultimately affect 1.5 billion people of the world population [40]. Land degradation process resulted in loss of productivity, depletion of greenery, loss of soil quality due to exposure of soil surface to wind erosion and water erosion. Data on irreversible land degradation is a serious threat for the conservation of agricultural land. About 5–6 million hectares of arable land worldwide are irreversibly lost each year due to soil erosion, salanisation and other degradation processes [41]. The process of chemical soil degradation is due mainly to excessive use of chemical fertilisers, pesticides and release of toxic industrial effluent into natural ecosystem without any treatment. The biological soil degradation is associated with depletion of soil organic matters and imbalance of soil nutrients. The Fifth International Scientific Agricultural Symposium held in 2014 suggests that land degradation has remarkable impact on the productivity or quality of cropland in some area, but its impact at regional and global levels is not yet clear. Enough evidences justified that land degradation affect all fronts of life [42–44]. The negative impact of land degradation is noticed to be severe in some regions and local areas due to a combination of resource factors (soil and precipitation) and economic factors (poverty, insecurity and lack of infrastructure) [45]. Response of farmers to overcome land degradation is either by adapting modern new farming technology or through independent and innovative farming technology [46].

Through Integrated soil fertility management (ISFM) practice by prevention mitigation, and restoration of land from desertification and land degradation can be managed by using balanced nutrients supply and minimum affective water management system for better productivity. By adapting IFSM some farmers have shown more than double production [47].

1.1.7.3 Climate Changes

The UN's Food and Agricultural Organization (FAO) in its report released on 12 September 2018 highlighted that climate change is one of the primary cause of rising global hunger. Climate changes badly affect production of major crops like rice and maize in tropical and temperate regions. The shortfalls in food availability lead to hike in food price and income losses that are responsible for reduction in people's access to food. Thus, in order to overcome the problem of hunger caused by climatic changes, the UN Food and Agriculture Organization (FAO), the International Fund for Agricultural Development (IFAD), the UN Children's Fund (UNICEF), the World Food Programme (WFP) and the World Health Organization (WHO) call for integrated approach to counter the adverse effects of climate change on food production system. The worst cases of extreme climate-related mishappenings like natural disasters, including extreme heat, droughts, floods and storms, have doubled since 1999s, with an average of 213 of these events occurring every year during the period of 1999–2016. The FAO also reports that the number of undernourished people tends to be higher in the countries experienced with extreme climatic changes. The WFP in collaboration with the United Kingdom has developed an interactive map that outlines different scenarios for the future. The Paris Agreement and 2030 Agenda for Sustainable Development have agreed the link between hunger and climate change as well as need for urgent action to protect the most vulnerable communities. For example, the food insecurity due to climate change is greater for Africa, where there is continued decline in agricultural production due to severe climatic change, in cyclic way [48, 49]. By 2030, the production of cereal crops will reduce from 15% to 19%, as the temperatures are estimated to rise from 1 to 2.75 °C which will result in less rainfall, and ultimately responsible for food insecurity in 2030 [49]. Extreme climatic change and desertification of agriculture land can be resulted in poor productivity and livelihood patterns, economic losses, and effect on infrastructure, markets and food security. The efficacy of food security is mainly depends on its link to our ability to adapt with the extreme events cause due to climate change.

1.1.7.4 Agriculture Diseases

Livestock and crops are prone to diseases, which may be ultimately responsible for net productivity. For example, Ug99, a lineage of wheat stem rust, can cause complete loss of wheat crop, if proper precaution is not taken, timely. Mainly, African and Middle East farmers have been suffering a lot due to this disease [50, 51]. In order to avoid such problem, genetically modified crop resistance to such disease will be helpful.

A vast number of pathogens bring disaster in crop cultivation. Generally, it is a herculean task to control plant pathogens because their populations are variable in time genotype. In order to save the crops from such attack, it is necessary to well understand the problems like proper identification of the pathogen and mechanism of action on the crop plants and seek remedies.

1.1.7.5 Food Versus Biofuel

Due to the depletion of natural fossil fuels, the demand of biologically derived green energy is gaining global popularity. Commodities like maize (corn) sugarcane and vegetable oil are being used as alternate to biological energy resources for liquid fuels. For example, since 2006, in the United States the agriculture land used for regular crop cultivation are also used for corn to get alcohol as liquid biofuel. The second generation biofuel like ethanol can be produced from maize and sugarcane and also meet the demand of food. Brazil is the world's largest sugarcane ethanol producer and pioneer in using ethanol as motor fuel. In 2015/2016, Brazilian ethanol production reached 30.23 billion litres (8 billion gallons). Brazil's sugar-based ethanol did not push food process appreciably higher [52, 53].

Use of farmland for agricultural production has long been a practice among farmers, mostly of developing nations. Farmlands are also being used for non-food crops like tobacco, opium, cotton, flax, rubber and fuel crops as fire wood. Since the twenty-first century the production of fuel crops has increased. However, technologies are also developed to commercially produce food from energy such as natural gas and electrical energy with tiny water and land foot print.

1.1.7.6 Food Waste

The Food and Agriculture Organization of the United Nation (FAO) reports that one third of the food production (1300 tonnes) is lost or wasted in developed countries. Poor food chain management systems are also one of the major factors in distribution and transport of food from its production sites to market. This prevailing poor food supply chain is responsible to create about 800 million hungry people at global level.

One of the best examples to prevent food waste is the initiative taken by the UN government to upgrade the lives of people of Kaduna, an agricultural region in northwest Nigeria. In Kaduna, the production of tomatoes and vegetables is vital to feed the population, as well as to generate employment and income. But due to the lack of adequate storage facilities and poor supply chain management system, as much as 50%–60% cereal grains can be lost after post harvest stage.

References

1. Zero Hunger Challenge (2012) the United Nations; www.un.org/en/zerohunger/challenge.shtml
2. Trade Reforms and Food Security Conceptualizing the Linkages (2003) FAO, UN. www.fao.org/3/a-y4671e.pdf
3. Ingram JC (2006) Bread and stones: leadership and the struggle to reform the United Nations world food Programme. Book Surge, N. Charleston. ISBN 141964470X
4. Loewenberg S (2007) Should the world food programme focus on development? Lancet 369(9580):2149–2150
5. Pisik B (2006) Sheeran Former Washington Times Editor, Will Lead U.N. Food Program. Washington Times National Weekly Edition November 13, 2006, 24
6. WFP (2013) An overview on world food programme English language essay. https://www.ukessays.com
7. Global Report on Food Crises (2018) World I ReliefWeb https://reliefweb.int/report/world/global-report-food-crises-2018
8. Monitoring food security in countries with conflict situations: A joint FAO/WFP update for the United Nations Security Council (2018). https://reliefweb.int
9. Venton C, Majumder S (2013) The economics of early response and resilience: lessons from Bangladesh. Department for International Development, London. Available at http://r4d.dfid.gov.uk/pdf/outputs/Hum_Response/61114_Bangladesh_Report.pdf
10. Horton S (2017) Economics of nutritional interventions. In: de Pee S et al (eds) Nutrition and health in a developing world, nutrition and health. Springer Science+Business Media, New York. 2017. https://doi.org/10.1007/978-3-319-43739-2_2
11. Strengthening resilience for food security and nutrition: a conceptual framework for collaboration and partnership among the Rome-based Agencies. April 2015. www.fao.org/resilience/resources/resources-detail/en/c/335336/
12. FAO, IFAD and WFP (2015) Strengthening resilience for food security and nutrition: a Rome-based agencies' conceptual
13. Framework for Collaboration and Partnership. Rome. https://docs.wfp.org/api/documents/WFP-0000062320/download/
14. FAO IFAD (2015) The state of food insecurity in the world 2015—meeting the 2015 international hunger targets: taking stock of uneven progress. FAO, Rome, p 56
15. Gustavsson J et al (2011) Global food losses and food waste. Food and Agriculture Organization of the United Nations, Rome
16. Green D (2013) The role of the state in empowering poor and excluded groups and individuals. United Nations Department of Economic and Social Affairs, New York. 2013 July
17. Nguyen HTM, Do H, Kay A, Kompas T, Nguyen CN, Tran CT (2017) The political economy of policy exceptionalism during economic transition: the case of rice
18. FAOSTAT. [http://faostat.fao.org/], []
19. UN: World Urbanization Prospects, the 2009 Revision (2010) Department of Economic and Social Affairs, Population Division, United Nations, New York
20. Morton JF (2007) The impact of climate change on smallholder and subsistence agriculture. Proc Natl Acad Sci U S A 104:19680–19680
21. Smith LC, Subandoro A (2007) Measuring food security using household expenditure surveys. International Food Policy Research Institute, Washington, DC
22. Headey D (2011) Rethinking the global food crisis: the role of trade shocks. Food Policy 36:136–146. https://doi.org/10.1016/j.foodpol.2010.10.003. View Article Google Scholar
23. FAO (2008) Briefing paper: hunger on the rise: soaring prices add 75 million people to global hunger rolls. Google Scholar, Rome
24. FAO (2010) Statistical yearbook. Food and Agriculture Organisation Google Scholar, Rome

25. Duke and Duchess of Cambridge visit UNICEF's life-saving emergency supply centre in Copenhagen to highlight desperate plight of children in East Africa. http://www.unicef.org/media/media_60370.html
26. Von Braun J, Ahmed A (2008) High food prices: the what, who, and how of proposed policy actions. International Food Policy Research Institute Google Scholar, Washington, DC
27. FAO, IFAD, IMF, OECD, UNCTAD, WFP, The-World-Bank, WTO, IFPRI, UNHLTF (2011) Price volatility in food and agricultural markets: policy responses. Food and Agriculture Organisation, Rome. Google Scholar
28. FAO: Statistical yearbook. 2010, Rome*:* Food and Agriculture Organisation Google Scholar
29. Matthew PR, Jensen AC, Gregory CA (2017) Understanding the prevalence, severity, and distribution of food insecurity in United States Statistic: Food & Nutrition Assistance http://www.healthpeople.gov
30. Hunger is a racial equity issues. https://moveforthunger.org
31. Committee on World Food Security (CFS), Global Strategic Framework for Food Security and Nutrition (2011). http://www.fao.org/fileadmin/templates/cfs/Docs1011/WG_GSF/GSF_annotated_outline_formatted_Rev1_22_Jun_11.pdf
32. UNICEF (1998) The state of the world's children, focus on nutrition, http://www.unicef.org/sowc98/silent4.htm
33. SIWI, IFPRI, IUCN, IWMI (2005) Let it reign: the NewfckLRWater paradigm for global food security, FinalfckLRReport to CSD-13. Stockholm International WaterfckLRInstitute, Stockholm
34. Klennert K (ed) (2009) Achieving food and nutrition security, Feldafingen/InWEnt
35. USAID (1995) Food aid and food security policy paper, http://transition.usaid.gov/policy/ads/200/foodsec/foodsec.pdf
36. IICA, Food Security (2009). http://www.iica.int/Esp/conocimiento/actualidad/Documents/Food%20security%20IICA%20Eng.pdf
37. Scherr SJ (1999) Poverty-environment interactions in agriculture: key factors and policy implications. United Nations development Programme and European Commission, New York
38. Beinroth FH, Eswaran H, Reich PF et al (1994) Land related stresses. In: Virmani SM, Katyal JC, Eswaran H, Abrol IP (eds) Stressed ecosystems and sustainable agriculture. Oxford and IBH, New Delhi
39. Eswaran H, Lal R, Reich PF (2001) Land degradation: an overview. In: Bridges EM, Hannam ID, Oldeman LR, Pening de Vries FWT, Scherr SJ, Sompatpanit S (eds) Responses to land degradation, Proc. 2nd. International Conference on Land Degradation and Desertification, Khon Kaen, Thailand. Oxford Press, New Delhi
40. Bai ZG, Dent DL, Olsson L et al (2008) Global assessment of land degradation and improvement. 1. Identification by remote sensing, Report 2008/01. ISRIC – World Soil Information, Wageningen
41. Scherr SJ (1999) Soil degradation: a threat to developing-country food security by 2020? UNCCD secretariat 2013. International Food Policy Research Institute, Washington, DC
42. Barrow CJ (1991) Land degradation: development and breakdown of terrestrial environments. Cambridge University Press, Cambridge, UK
43. Blaikie P, Brookfield H (1987) Land degradation and society. Methuen, London
44. Johnson DL, Lewis LA (1995) Land degradation: creation and destruction. Blackwell, Oxford
45. Wiebe K (2003) Linking land quality, agricultural productivity, and food security, resource economics division, economic research service, U.S. Department of Agriculture. Agricultural Economic Report No. 823. www.odi.org.uk/resources/download/4662.pdf. Accessed on 28 Jan 2010. View publication stats
46. Mermut AR, Eswaran H (1997) Opportunities for soil science in a milieu of reduced funds. Can J Soil Sci 77:1–7
47. Divyalakshme A, Divyagopalakrishnan I, Nivethaa K et al (2013) The analysis and assessment of land degradation. Int J Appl Eng Res 8(16):1923–1928
48. Brown ME, Funk CC (2008) Food security under climate change. Science 319(5863):580–581
49. Devereux S, Edwards J (2009) Climate change and food security. IDS Bull 35(3):22–30

50. McKie R, Rice X (2007) Millions face famine as crop disease rages. The Guardian. UK April 22, 2007. Retrieved 13 Nov 2011
51. Mackenzie DB (2007) Billions at risk from wheat super-blight. New Scientist April 3, 2007. (2598): 6–7. Retrieved 2007-04-19
52. Veja Magazine (2008). Etanol não influenciou nos preços dos alimentos (in Portuguese). Editora Abril. Archived from the original on 2016-05-19
53. Biofuels major driver of food price rise-World Bank. Reuters. 2008-07-28. Retrieved 2008-07-29
54. Shaw DJ (2007) World food security: a history since 1945. Palgrave Macmillan, Basingstoke/Hampshire/New York
55. Boyd Orr J (n.d.) As I recall (London: McGibbon and Kee, 1966) and M. Gowing, 'Richard Morris Titmuss'. Proc Br Acad 61(1975):1–30
56. The Hot Springs Conference (1943) FAO www.fao.org/docrep/009/p4228e/P4228E04.htm
57. The Food and Agriculture Organization of the United Nations (1947) International Organization; Vol. 1, No. 1 (Feb., 1947), pp. 121–123; https://www.jstor.org
58. Simon G- A (2012) Basic reading as an introduction to Food Security. Int J Humanit Soc Sci Roma Tre, 2012, VI(IV)
59. Bhawmick S (2018) Food security in India: challenges and prospects. Pratidhwani Echo Peer-Rev Int J Humanit Soc Sci VI(IV):307–329
60. "Background to the Voluntary Guidelines" (2012) The voluntary Guidelines, Food and Agriculture Organization of the United Nations, www.fao.org/righttofood/vg/about_en.htm. Accessed 21 May 2012

Water, Energy and Food Security: Pillars for Zero Hunger

2

2.1 Introduction

The earth-like habitable planet needs water, energy and food for the survival of life. Water is the primary source of survival of living beings. Thus, water security concerns the most vital factor for the sustainability of life. Life means activity, and for activity energy is inescapable. Energy conservation plays a complementary role in survival of life, as water. Thus, security of water and energy at different geographical locations is in foremost priority to meet the challenge of global population, which is supposed to reach 9.6 billion by 2050, with the trend of water demand projected to increase by 55%, energy demand supposed to enhance by 80% and food demand projected to increase by 60%. Thus, Rio+20 declaration emphasises on water, energy and food security which may ultimately be helpful while integrating economic, social and environmental factors to eradicate hunger and malnutrition crisis, at global level. Water-energy-food nexus is prevailing all around us, but is ignored by the public, governance and policymakers. The water-energy-food nexus can be best cited while explaining significance of water for all sort of activities. For example, energy is needed to treat (purify), distribute (pump) and boil water for thermal electric generation and running of hydropower, and water is an essential component of our food systems through the irrigation of crops and animal consumption. But, the nexus is not restricted to microlevel; it can be widely implemented in projects like land-use planning, human health management and environment protection. The drive for nexus is just not due to interdependent relationships between water, energy and food but also depends on the dynamic nature of climate like rise and fall of temperature, pressure, rainfall and natural calamity and even factors like the socio-economical condition prevailing in a state and the nature of governance. These three sectors (water, energy and food security nexus) are key driving forces to eradicate hunger and to bring sustainable development.

B. K. Behera et al., *Move Towards Zero Hunger*,
https://doi.org/10.1007/978-981-32-9800-2_2

2.1.1 Water Security for Food

"World Water Day" has been celebrated since 1993 in order to divert the attention of the people, governments, international organisation and voluntary bodies on various issues related to water conservation, management and utilisation. Drive for clean and safe water, sanitation, hygiene and use in the agricultural and energy sector has been carried out to challenge the water sector to overcome the problems of pollution of freshwater bodies and groundwater. So, proper acquisition of water resources and management, for growing crops and raising livestock and use in day-to-day works, is linked to food safety and conservation. Timely implementing water security and conservation programmes can only be possible if everyone is involved: from policymakers, water sector workers, farmers and teaching community. Thus agriculture water management is the key for food production and food security. Inadequate water management practice may lead to depletion and degradation of land and water resources. So, it is utmost necessary to link both water security and food security while protecting and improving aquatic ecosystems.

Availability of sufficient food in rural areas mainly depends on access to natural resources, including water for agriculture. In 2010, the UN General Assembly declared access to clean drinking water and sanitation as a human right. But the right to water in context of the right to food is unsolved. Although, enough water is available for our future needs, but, due to improper management, billion of people have been facing problem of water scarcity [1, 2]. Major changes in government policy and management, across the entire agricultural production chain, are needed to ensure best use of available water resources to bring sustainability in food productivity and to manage food security [3, 4].

2.1.1.1 Water Scarcity

All forms of life need water for survival. About three-fourths of the earth's surface is occupied by water. But about 96.5% of the global water resources are of oceans and seas origin, and out of which 2.5% is useable freshwater (Fig. 2.1). The volume

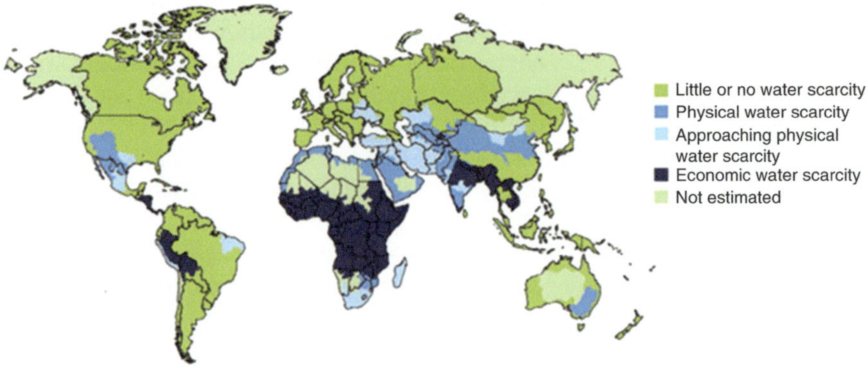

Fig. 2.1 Global water scarcity. (Source: Wikipedia 2019; [2])

of freshwater in ice sheets and glaciers is about 70%, and stored groundwater is 30%. Deficit of freshwater resources to fulfil the minimum requirement of common people is known as water scarcity. In 2015, the World Economic Forum announced the global risk for water would be faced within the coming decade [5].

It is mainly caused by limited availability of quantity and quality of freshwater, dispute between the users (at both state and country levels), irreversible depletion of groundwater and negative impact on the environment. One-third of the global population (2 billion people) has been suffering from severe water scarcity [6–9].

Only 0.014% of all water on earth is apposite for drinking and other household purposes, and of the remaining 97%, water is either saline or less than 3% hard to access (Fig. 2.2). Sharp, increase in urbanisation and industrialisation have been acting as catalyst factor for water crisis.

Water scarcity is mainly due to human interference with water cycle. In addition, the economic policy, planning and management of government are also responsible for inadequate availability of consumable water for food and survival of life.

2.1.1.2 Supply and Demand of Freshwater

It has been estimated that about 14,000 km^3 (nearly 3350 mi^3) of fresh surface waterbodies is accessible for human use. But, only 5000 km^3 is being used and reused by humans. So, it is felt that more freshwater should be available to meet the demand of population of 8 billion, by 2020. But, due to unequal geographical distribution with variable rate of water consumption, it is a scarce resource in some parts of the world and some parts of the population. The poor supply of water may be due to two reasons, i.e. physical water scarcity and economic water scarcity. The former

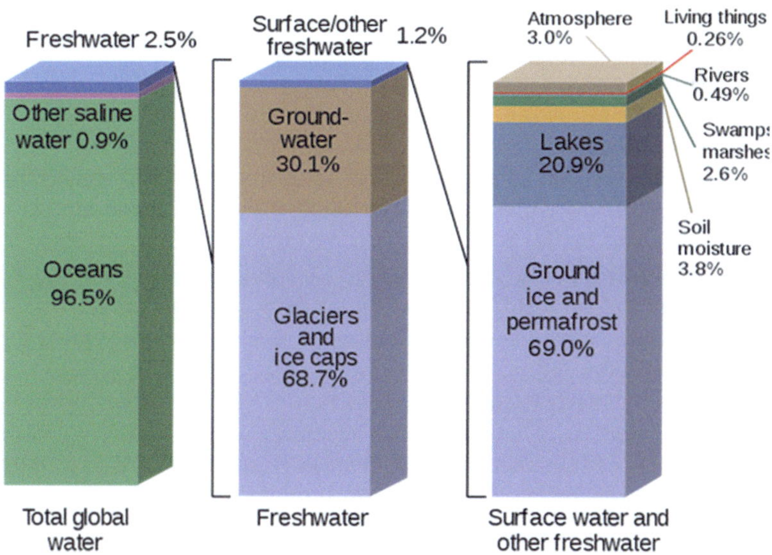

Fig. 2.2 Status of waterbodies on earth. (Source: Wikipedia, 2019)

problem results from inadequate natural water resources for water supply chain, on a demand basis. The latter problem resulted from mismanagement of water resources to make them available to public, timely. In order to overcome such problem, the United Nations Committee on Economic, Social and Cultural Rights established a foundation of five core attributes for water security. These are mainly based on (a) insufficient access to safe drinking water, (b) mismanagement for sanitation and waste disposal, (c) excessive pumping of groundwater, (d) unstable biodiversity due to water pollution and (e) dispute over the use of water resources. The basic target of such organisation is to strengthen human right for sufficient, safe, acceptable, physically accessible and affordable water for personal and domestic use. That is why in the 2030 Agenda for Sustainable Development, maximum emphasis is given for safe drinking water, effective sanitation and good hygiene (WASH). So, the action plan taken for the drive for safe drinking water is to be reviewed, time and again, to meet the challenge at the extreme geographical location where water scarcity prevails due to both physical water scarcity and economical water scarcity. It is about three decades back; in 1990, WHO/UNICEF Joint Monitoring Programme for Water Supply, Sanitation and Hygiene (JMP) had a thorough survey on the progress status of drinking water, sanitation and hygiene (WASH) at global front. In 2009, JMP established a Strategic Advisory Group (SAG) to provide independent advice on the continued development of the JMP as a custodian of global data on drinking water, sanitation and hygiene. Since then, the JMP has been keen on keeping updated information on various issues on availability of safe drinking water, especially in rural location where both the physical scarcity and economical scarcity of freshwater are prevailing. The JMP 5-year strategy (2016–2020) focuses on further enhancing global monitoring of drinking water, sanitation and hygiene (WASH), keeping in view of 2030 Agenda, and to convey it to decision-makers at government and civil society organisation level.

2.1.1.3 Water Availability in Rural Areas

2.1.1.3.1 Groundwater

Groundwater is the primary source of water in rural areas. Among the several water supply systems, groundwater is far the most practical and safe in nature. The advantages of groundwater are as follows: (a) it is free from pathogenic microbial contamination, (b) it does not need any physic-chemical treatment, (c) it can be available in the close vicinity of rural areas, and (d) it is economical to maintain. However, the groundwater is often contaminated by excess of minerals, which may cause harmful effect on human health.

Generally, rainfall percolated into the earth to form underground deposits known as aquifers, which are used without further treatment for individual or community level in rural areas or free-water zone. Wells can be drilled into the aquifers, and water can be pumped out for use. The underground water table can eventually change over time due to change in weather cycles and precipitation patterns, streamflow and geologic changes or human-induced changes, like the increase in impervious surfaces on the landscape.

In case of non-availability of conventional electricity, windmills are used to generate electricity for underground water pumping. The windmill's turbine blades capture energy from the wind and turn it into mechanical energy by spinning a generator that creates electricity. Like windmill, solar technology is a good source of nonconventional energy, especially in rural areas where conventional electricity operation is extremely difficult. This is a one-time investment for farmers, and once the costs are recovered, fuel is free. This would immensely be beneficial to farmers to save money and easy to maintain. Besides pumping groundwater, the solar panel-based electricity can be used for biomass processing and grain separation and polishing, even at the site of production.

2.1.1.3.2 Harnessing Seawater for Rural Livelihood

The coastal zone is a transition area between marine and territorial zones. It is necessary to ensure livelihood security of the fishing communities as well as other local communities to provide them safe water and hygienic food for a sustainable pattern of normal livelihood. Harnessing seawater by desalination is an alternate way and means to produce safe drinking water. In addition, seawater can be used for agriculture, after proper amendment. For example, in 2016, a UK-based agriculture company, named Sundrop Farms Holdings Ltd., has developed a greenhouse for tomato farming in the coastal area of Australia (Port Augusta). The facility runs on energy mostly produced by a 115 m solar tower that draws sunlight from 23,000 mirrors surrounding it.

Presently, the global water desalination process is in high demand with an annual growth rate (CAGR) of 7.8% from 2018 to 2025. It has been estimated that about 1% of total demand of freshwater is via desalination process.

The desalination process of seawater is carried out by means of thermal distillation or by membrane separation. Thermal distillation involves heat for boiling seawater into vapour phase, which is condensed back into water by cooling it down. The entire process is tedious and cost-effective. Currently, the most common process is membrane separation, in which semipermeable membrane is used. This process is known as reversible osmotic process, which is less expensive than thermal conversion process.

2.1.1.3.3 Glaciers as Water Source

Glaciers are the potential sources of water due to their contribution to streamflow. Due to rise in global temperatures caused by greenhouse effect, glaciers get melted, causing glaciers in general to shrink worldwide. Gradual shrinking of glaciers in the long run will diminish the available water resources. Due to water crisis, scientists and engineers are keen to exploit freshwater in icebergs by towing them to where ice can be melted and converting them to drinking water. Three-fourths of the freshwater are held in the form of iceberg, mostly covered in the Antarctic. The most dilemma and challenge issues, presently being faced by scientists and engineers are how to convert the terminal icebergs into fresh water. It is estimated that waste heat from electrical power can be used for melting icebergs. It could be possible to use waste heat from 10,000 MW power plant for melting 1,000,000 acre-feet of water-equivalent ice in 1 year.

About 100 "iceberg-producing glaciers" are in North Atlantic, of which about 40,000 medium- to large-sized icebergs break annually. Some of the icebergs are about 15,000 years old. Technically, icebergs are pure freshwater having salt cover on their surface.

2.1.1.3.4 Rainwater Harvesting

Rainwater harvesting is also a better option for rural areas lacking surface water or water reservoir systems. The rainwater harvesting practice can be well adapted in countries where rainfall is heavy, with long intervals with no rainfall. It is also a suitable source of water in arid and semiarid areas where people live in scattered settlements and no other sources of water reservoirs are available. Rainwater harvesting works well for household- as well as community-level supplies. Rainwater harvesting at household level can be managed by storage of rainwater through roof catchment and at community level by storage through ground catchment.

2.1.2 Energy Security for Food

Energy security and sustainable food availability complement each other. So, proper energy management is an unavoidable practice to reduce poverty, as the targets set out in the Millennium Development Goals (MDGs) are to be met. Almost 1.1 billion people have limited access to modern energy services for heating and cooking (IEA) [10], and 1 billion have zero or limited access to electricity, as reported by Derisking Renewable Energy Investment (DREI) [11]. Without access to electricity and sustainable energy sources, communities have little chance to achieve food security and no opportunities for securing productive livelihoods that can lift them out of poverty.

Currently, a wider gap is left between energy demands and energy access. It could have been lack of proper coordination between the ruling government and the various international volunteer bodies and other organisations to implement policy decision related to conventional energy supply network or development of nonconventional energy projects. Long back, in apprehension, the United Nations Sustainable Development expressed desired for bringing profound amendment in energy supply in agriculture system for 2 billion people expected to be under nourished by 2015 [12].

Energy crisis in many regions is a fundamental barrier to minimise hunger and to meet the future demand of food. It is not practicable to resolve food crisis without massive implementation of nonconventional means of energy production, primarily in rural locality of developing nations. From a household perspective, access to modern energy services is still extremely problematic in many developing countries.

2.1.2.1 What Is Energy Security?

The IEA defines energy security as "the uninterrupted availability of energy sources at an affordable price". In a little elaborative form, energy security means that energy sources are sufficient to meet the energy demands of a political community, which include social, economic and military activity, and that this demand will be met in a reliable, stable manner in the future [13]. There are various degrees of energy security with differing consequences for countries. In general, when demand is not met, citizens' daily needs, it leads to problems like food insecurity, healthcare, education and transportation, other day-to-day meets for a normal livelihood [14]. On a much larger scale, countries can be affected militarily and economically.

The energy demand for agriculture practice, food storage and food processing can be met in the form of two dimensions: (i) long-term energy security and (ii) short-term energy security.

In the former case, it deals with timely investments to provide energy in line with economic development and sustainable environmental needs to protect crops and aquatic food under normal or any adverse conditions caused by climatic changes or natural calamity. The long-term measures in energy security mainly make a nation free from or dependent on imported energy and more reliant on finding out native renewable resources for sustainable energy. A nation can strengthen energy security by monitoring energy demand through conservation and developing social awareness, for the susceptibility of natural renewable resources. It can also involve entering into international agreements to underpin international energy trading relationships, such as the Energy Charter Treaty in Europe.

In the latter case, short-term energy security focuses on the ability of the energy system to react promptly to sudden changes within the supply-demand balance. Lack of energy security is thus linked to the negative economic and social impacts of either physical unavailability of energy or prices that are not competitive or are overly volatile.

Energy is power, which is unavoidable for human activity in this habitable world. Thus, conservation of energy resources at different geographical locations is a primary issue for a government to support and enhance agricultural productivity, adequately available to the people, even staying at remote places. From political, economic and environmental fronts, energy security is one of the most important issues faced by all countries in the world. As such, energy has a fundamental role in states' structure, consolidation and survival. Besides this, energy is an important tool to drive the SDGs goal scheduled for 2030. Considering the competition between states, energy is a crucial factor in the distribution of world power. Therefore, those countries with the most control of energy resources have the biggest power advantage in the international system and high productivity.

Today, energy security is an important political issue due to the rapid industrialisation of the world, growing populations, high levels of consumption, unusual climatic change and a significant dependence on non-renewable fossil fuels. Major powers are going to great lengths to establish and guarantee their energy supplies. More and more, they are militarising their approach to energy security, as evidenced by the United States' involvement in the Iraqi invasion of Kuwait in the 1990s, the 2003 invasion of Iraq and the subsequent removal of then dictator Saddam Hussein.

National security and availability of energy resources are the two important aspects of a country and depend on how the government and associated voluntary organisations exercise their power and administrative ability to form and operate energy network. Access to inexpensive energy has become essential to the functioning of modern economies. However, the uneven distribution of energy supplies among countries has led to significant vulnerabilities. So, implementation of energy security policy in an effective way has been an increasingly debatable issue for policymakers, entrepreneurs and academics who usually claim to pursue it when proposing or implementing changes in the energy domain. Still, we are in dilemma as it is often not clear what is the precise meaning of energy security, especially when approaching it from an economic perspective. The relationship between food security, economic growth, human well-being and energy security is interlinked to each other and can be captivated in multiple respects. This can be well operated under the guidelines of SDGs targeting to eradicate the discrimination in energy distribution by the end of 2030.

Unusual climatic changes, environmental pollution, politicisation of the *bureaucracy*, *natural energy resources depletion* and the growing dominance of market forces all have profound implications for energy security, which is ultimately affecting food productivity and distribution. In fact, the past and present of global energy scenario explain that the increase in monopolisation of non-renewable energy resources and *politicisation* of *bureaucracy* in energy policy decision by resource-rich countries have made energy security a matter of national security [15]. Factually, energy security needs global collaboration, government involvement and even legal action. Energy security does not exclusively mean the exploitation of both the conventional and renewable energy resources, but it is foremost necessary to monitor the energy supply chain, so that it should meet the requirement of remote people to manage agriculture activities and other livelihood purposes for a sustainable normal life. It has been estimated that $38 trillion dollars needs to be invested in the energy supply infrastructure between 2011 and 2035 to meet the growing demand [16].

2.1.2.2 Access to Energy

SDG 7 aims to correct the wider gap between energy demand and access, by ensuring availability of electricity to everyone as affordable, reliable and modern energy services by the year 2030.

2.1.2.2.1 Access to Electricity

Availability of electricity to the common people of a particular area is otherwise known as electricity access or electrification rate. Electricity access plays as an indicator of developing status of a nation. It has been surveyed that about 1.2 billion people are deprived of electricity in their home, many of these, about 95%, being from Africa and Asia [17]. About 84% people, worldwide, lack access to electricity [18]. Recent report says that over 120 million people, globally, have access to electricity, in 2017 [19].

The developing countries like India have already covered electrification of almost all rural areas by 2018. Many Asian countries have also dawn significance progress in electrification. In Indonesia, 95% of habitable area has been facilitated with electricity, whereas Bangladesh has covered about 80% electrification by 2018.

Meanwhile, under developing countries in Africa, Kenya's access rate to electricity has progressed significantly from 8% in 2000 to 73% by 2018. In Ethiopia, the electricity facility has covered about 45% of total locality of the state. The National Electrification Program has set out a plan to reach universal access by 2025, aiming to reach 35% of the population with off-grid solution.

2.1.2.2.2 Access to Natural Gas

The most densely populated countries like China and India are in the process of replacing natural gas with coal. In China, currently, natural gas makes up about 6% of the country's total energy. However, according to China's gas-to-power plan, natural gas market share in the country will reach to 10% of the total energy consumption by 2020, and this will reach 15% by the end of 2030. In 2018, natural gas production in India has increased from 0.6% to 2.7 billion cubic metres (bcm) [20]. In china, the national data highlights that about 14% reduction in the practice of using of firewood was reduced significantly, during the period 2011 to 2015.

In Sudan, around half of the population now uses domestically refined LPG for cooking. In Cameron, the use of LPG has reached about 20% of the total population and will reach the target of 58% by the end of 2030. Kenya LPG is now used by 24% of urban household. Due to increased awareness for health, most of the women prefer to use natural gas for cooking and other household activities. Even people from rural areas have started using natural gas or biogas, instead of direct firing biomass for gating energy. For example, in Myanmar the government and policymakers are motivated to promote clean cooking in part because of the high rate of deforestation for firewood collection. As a result, reliance on biomass has decreased from 94% of the population in 2009 to 76% in 2015, mainly replaced by electricity. Ethiopia is one of the least developed countries in the world, but, in the recent past, use of electricity as a part of cooking practice has been gaining popularity. Approximately 34% of its over 100 million inhabitants live below poverty line [21]. The people of Ethiopia mostly depend on biomass as the source of energy followed by oil (5.7%) and hydropower (1.6%) [22].

In spite of all these efforts and progress, the world still remains off-track in its efforts to achieve Sustainable Development Goal (SDG) 7.1 to ensure universal access to affordable, reliable and modern energy services by 2030. Still, 2.7 billion people lack access to clean cooking facilities worldwide, relying instead on biomass, coal or kerosene as their primary cooking fuel [23].

Since the last two decades, the IEA has been in the process of serving SDG 7.2 agenda for renewable energy and 7.3 agenda on energy efficiency. In addition IEA is in touch with the other associates (IRENA, UN Statistics Division, WHO and World Bank) of SDG 7 campaign for the next milestone in global access to modern energy for all people worldwide.

2.1.2.2.3 Access to Renewable Energy

It is advisable to have a variety of options for energy in rural areas where extreme climatic condition prevails and conventional pattern of electricity is not penetrable. Under such condition renewable energy is a better option to meet the demand for agriculture practices. Energy from renewable resources like wind, water, solar, biomass and geothermal is inexhaustible and clean. Although the solution to energy's climate crisis lies off-grid, renewable energy currently constitutes only 15% of the global energy mix. So, it is high time for a new global partnership on sustainable energy for all, guided by SDG 7 on universally accessible, efficient, clean and reliable energy sources and services.

2.1.2.3 Solutions to Energy Crisis in Rural Areas

The sources of energy availability in rural areas are in the form of conventional energy (coal, oil, petrol, natural gas, electricity, coke) or locally accessible renewable energy (biogas, waste biomass, firewood, solar panel, agriculture residues, water and wind). In rural areas, energy is mainly used for agriculture practice, households, rural domestic animal farming and transport.

Still, rural families use traditional fuel (firewood, dry cow dung, biogas from fresh cow dung, agricultural residues) as a part of energy source (Fig. 2.3) due to extreme poverty. A major portion of which is collected by private effort, mostly from forest or postharvested field (dry stock of crops). The bulk of the fuel

Fig. 2.3 Energy source of rural people where no access to electricity. (**a**) Use of firewood for cooking, (**b**) use of cow dung for cooking, (**c**) use of biogas for cooking and other household works, (**d**) firewood collected from forest

requirement in the households sector is for cooking. Due to gradual depletion in availability of firewood, even traditional fuels are getting commercialised.

Thus, to overcome this prevailing rural energy crisis, it is high time to plan for sustainable development, as proposed by various international bodies, time to time. The rural areas of Africa, Asia and Latin America meet 75% of their energy demand from non-commercial energy sources like firewood, cow dung and agricultural wastes. The economic condition of rural people could be improved by increasing agricultural and nonagricultural income-generating activities. But, for this purpose, provision of subsidiary for conventional fuels like petrol, diesel and coal should be provided to the economically weaker farmer communities and to the working groups. In addition, the government should initiate and promote the use of local renewable energy by establishing proper training programme on basic operation protocol on tools and techniques related to renewable energy utilisation.

Farmers and labours are the key driving forces to support agricultural productivity. Thus, in 2018 Human Rights Council, the UN General Assembly passed a declaration on the rights of peasants and other working groups in rural areas. This was mainly based on the studies by the Advisory Committee, which had identified five main reasons for the disproportionately high share of hunger in rural populations. The main five causes include (i) expropriation of land, forced eviction and displacement, (ii) gender discrimination, (iii) absence of agrarian reform and rural development policy, (iv) lack of minimum wages and social protection and (v) criminalisation of movements defending the rights of people working in rural areas. So, for the implementation of any sustainable rural development programme, it is first necessary to have proper review and amendments in existing policies and programme, adequate budgetary provision and establishment of effective network for sustainable development. So, many sister concerns (FAO) of UN are involved to facilitate coordination between energy and environment and to promote the concept of sustainable development.

Keeping in view the major demand for fuel wood, FAO, in association with other UN agencies, has started several programmes to develop fuel wood supply network in rural areas of the developing countries, for example, through its Tropical Forestry Action Plan. In association with various bilateral and multilateral agencies, several developing countries have started social forestry programmes with the locally available resources. In addition, these agencies are also keen in being involved with monitoring the adverse effect caused due to unusual climatic changes, destruction of forest and urbanisation of cultivated lands in developing nations. The impacts of environmental awareness are already being observed by the blocking of ongoing power projects by environmental agencies and groups in some major developing countries.

So, it is high time to have vigilance over energy supply chain in developing nations, and more emphasis should be given to the rural energy situation through deliberate intervention by national government and international agencies, in order to increase the energy supply from all types of energy resources.

Sustainable pattern of rural development could only be possible when energy needs are integrated with agricultural activities and environmental conservation at all levels of geographical locations with proper planning network. In connection with this, FAO has been planning to develop such framework for integrated rural energy planning and assessment in order to meet the energy requirements for sustainable agriculture and rural development, with the lowest possible cost to the economy and the environment.

2.1.2.3.1 Move for Renewable Resources

The best possible solution is to reduce the world's dependence on non-renewable resources and to improve overall conservation efforts. Fossil fuel consumption (coal, oil, gas) have been playing key to drive Industrial Revolution, since 1750, and are responsible for social and economic development of a country. But due to the increase in population magnitude and activities, industrialisation is rapidly intensifying. The densification of industrialisation causes alarming problems such as air, aquatic and land pollution and depletion of ozone layer mainly caused by manufacture of chemicals, especially manufactured halocarbon refrigerators, solvents, propellants and foam-blowing agents (chlorofluorocarbons (CFCs), HCFCs, halons), referred to as ozone-depleting substances (ODS).

Currently, many alternate technologies are available for using renewable energies such as solar, wind and biogas. The world must therefore balance the role of energy in social and economic development with the need to decarbonise, reduce our reliance on fossil fuels and transition towards lower carbon sources. The major concern is not so much that we will run out of gas or oil but how to overcome environmental problems with the supplement of renewable energy, in a balanced manner.

Biogas is a better and eco-friendly source of energy. The basic substrates for biogas generation include cow dung, agricultural wastes, industrial organic wastes, night soil, waste from animal slaughter house and a wide range of organic wastes generated from human activities. Some of the organic wastes are abundant in rural areas and available freely. Biogas is produced by anaerobic digestion of these organic wastes. Currently, the tools and techniques for biogas are easily available from the government or voluntary organisations. The composition of biogas includes methane (60–80%) and carbon dioxide (20–40%) with trace amount of hydrogen sulphide, nitrogen and other impurities. Well-established standard protocols are available in various agencies dealing with renewable energy. Biogas for electricity production or compressed liquid gas has been widely used for both industry and household works (Fig. 2.4). After biogas production, the leftover slurry can be used as a potent organic fertiliser due to its high mineral contents. The slurry also upgrades the structure and water retention capacity of the soil.

Wind energy is also a better option over the conventional fuels. Running of wind turbines is free from the application of electricity or liquid fuels. It is also a pollution-free source and can be installed in the coastal area and at the top of rounded hills, open planes and gaps in mountains (where wind is strong and reliable). Conventional power plants operate at the cost of heavy water to generate power from fossil fuels,

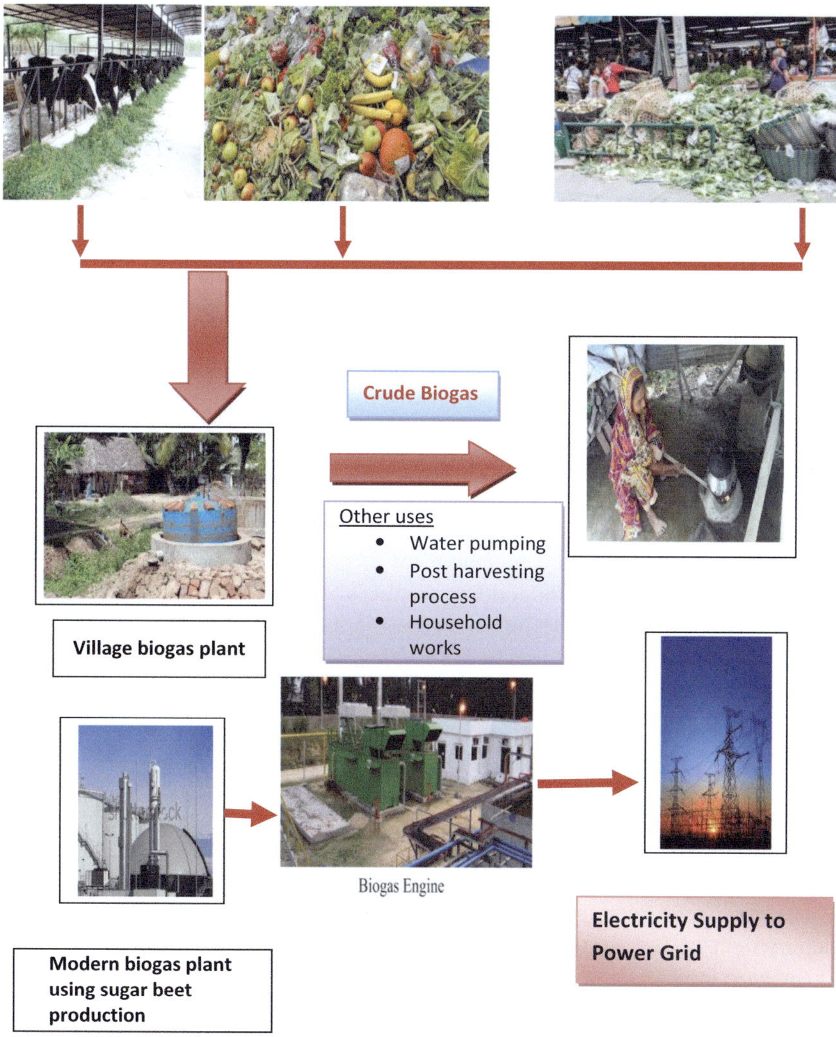

Fig. 2.4 Biogas production and utilisation network in rural and urbanised areas

whereas wind turbine is free from such energy input. Thus, drought-stricken locality in rural America uses wind energy for agricultural practices.

In general, to meet energy demand is expensive, and the management for energy distribution network is quite tedious and highly cost-effective. The use of solar power, specifically in rural areas, is a cheaper and cleaner alternative with minimum financial involvement. Solar energy use is beneficial for environment management programme, due to the zero emission of any harmful gas. India which is a developing country has the world's biggest solar panel system, Kamuthi Solar Power Station in Tamil Nadu, having energy generation capacity of 648 MW covering 2,500 acres and consisting of 2.5 million solar panels. The site is estimated to make enough power for 750,000 people.

2.1.2.3.2 Buy Energy-Efficient Products

It is usual practice of any developing country to supply adequate electricity to urbanised area while ignoring the minimum energy requirement for rural locality. This uneven distribution of electricity fails to meet the demand of electric energy for agricultural activities such as postharvesting process and management of cool tower for the storage of perishable agriculture products and to meet the energy demand for regular household works and operating village agro-industry. So, it is necessary to reduce electricity expenses of urbanised areas by using energy-efficient products like replacing traditional bulbs with CFLs and LEDs. These electrical appliances consume lesser watts of electricity and save lots of power, which can be diverted to rural locality, nearby.

2.1.2.3.3 Lighting Controls

There are a number of new technologies out there that make lighting controls much more interesting, and they help to save a lot of energy and cash in the long run. Present lighting controls, slide lighting controls, touch dimmers and integrated lighting controls are few of the lighting controls that can help conserve energy in urbanised areas. This practice may be helpful in saving lots of energy and diverting it to rural villages.

2.1.2.3.4 Grid Access

Grid extension to rural areas may be possible, but would be not economical due to the uneven distribution of low-density population in rural villages and low electricity consumption in lower-income households. As high-power grid connection costs heavily, rural people are not economically sound enough to bear the electricity cost, without any subsidy scheme from the government. So, high cost can be a barrier to the deployment of decentralised off-grid system. So, the government should have well provision for plug into the grid and getting credit for power on need basis.

2.1.2.3.5 Energy Simulation

Currently, due to lack of awareness on renewable energy, a large number of rural families are using fossil fuels for household works. The unavailability of adequate conventional energy deprives the rural people from the basic needs such as lighting and communication but also limits productivity and economic development in this area. So, the government and nongovernmental organisations (NGO) are in touch with the rural villages to provide provision for modern renewable energy sources. Although electricity access is essential for rural development, it may not be alone linked to poverty eradication and food crisis. Integration of energy supply network with economic planning and proper implementation could improve the standard of rural people. Proper efficient energy simulation network should be developed to link urbanised area with rural energy supply system for the better use of surplus energy being saved from urban areas. Joint efforts of engineers, architects and designers could develop software simulation for the management of rural energy supply system.

2.1.2.3.6 Energy Audit

Detail analysis of energy accessibility at a place, nature of energy resources, detail on capital expenditures (CAPEX), final cost of energy at user level is known as energy audit process. This process involves balancing the total energy inputs with its use and serves to identify all the energy streams in the systems and quantifies energy according to its discrete function. In addition, energy audit explains about energy cost optimisation under extreme circumstances of locality safety aspects and also gives clues to improve the operating and maintenance practices of the system. So, irrespective of location of a place, energy audit is an effective major for both conventional and renewable energy systems. It is advised to carry out energy audit through a professional team who can assist in reducing carbon footprint, save energy and unnecessary expenses and avoid energy crisis.

2.1.2.3.7 Common Stand on Climate Change

Both developed and developing countries should adopt a common stand on climate change. They should focus on reducing greenhouse gas emissions through an effective cross-border mechanism. With current population growth and overconsumption of resources, the consequences of global warming and climate change cannot be ruled out. Both developed and developing countries must focus on emissions cuts to cut their emission levels to half from current levels by 2050.

2.1.2.3.8 Biowaste Recycling for Methanol Production

Biomethanol can be produced from a wide range of biomass feedstocks via a thermochemical route similar to the Fischer-Tropsch process for BtL. It can be blended in petrol at 10–20%. In China, M10 and M85 are already used in thousands of vehicles. Methanol has also been investigated for use as a fuel in shipping. BioMCN, Netherlands, is the first company in the world to produce market and sell industrial quantities of biomethanol, using glycerine as a feedstock. BioMCN continues to develop innovative processes for the production of biomethanol using various feedstock, including crude glycerine, green gas, biomass and CO_2. Presently, USA is also showing keen interest in bio-methanol production from garbage having rich in: organic wastes, food wastes, waste papers, scrap wood wastes.

2.1.2.3.9 Use of "Flex-Fuel" in Automobiles

Flexible-fuel vehicles (FFVs) are designed to run on a certain blend of ethanol with unleaded gasoline with concentrations of ethanol as high as 85 percent, thus giving us the E85 derivative. More than 17 million vehicles in the United States come off the assembly line as flex-fuel, meaning they can run perfectly well on any mixture of gasoline and ethanol, up to E85 (which is actually 51–83% ethanol, the rest gasoline).

In the United States, the gas-powered cars were having average street life of 17 years. It means that the car owners have to depend on compressed liquid petroleum gas for about a period of two decades. So, in order to provide better option, the United States made it mandatory for car manufacturers to produce flex-fuel technology-based cars where any hydrocarbon fuels like gasoline, ethanol, methanol, etc. can be used.

Running of flexible-fuel vehicles can save lots of oil and petrol. The surplus amount of fossil fuel saved can be diverted to rural villages for running transport system linked to agricultural activities like running water pump, maintaining cold storage houses and operating machineries for postharvesting purposes.

2.1.3 Food Security

The United Nations Committee defines food security as "All people, at all times, have physical, social, and economic access to sufficient, safe, and nutritious food that meets their food preferences and dietary needs for an active and healthy life" (Fig. 2.5).

2.1.3.1 Factors for Food Availability

The four main factors for food security are food availability, access to food, utilisation and stability. All these factors are interlinked, and the cumulative efforts serve to manage sustainable food security for a nation (Fig. 2.6).

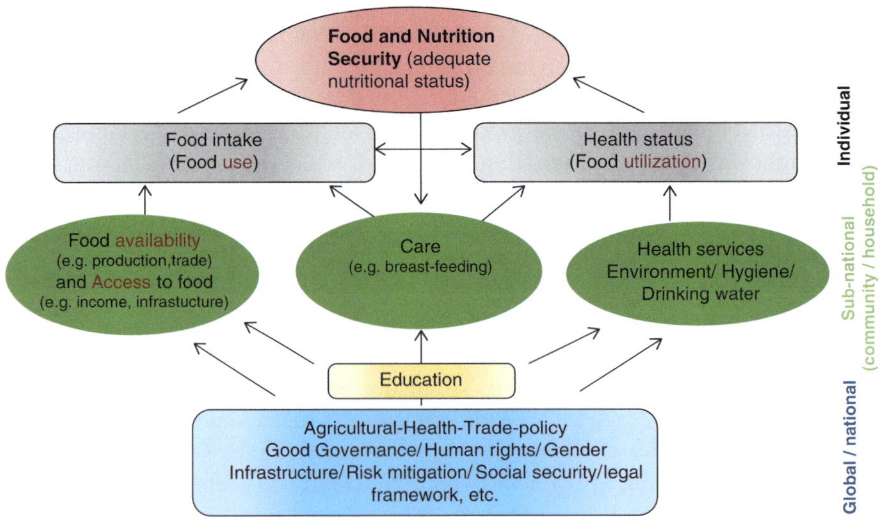

Fig. 2.5 The graph shows the complex aspects and interplay of food and nutrition security on the different levels of social front. (Source: modified after UNICEF [24])

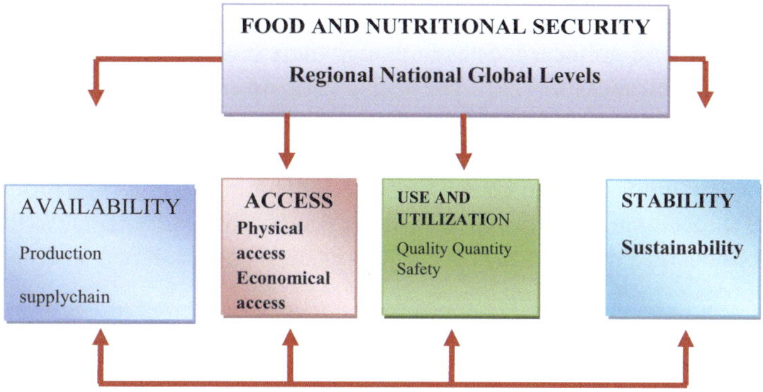

Fig. 2.6 Food and nutritional security network

Different other factors like agriculture production, existing local resources for renewable energy, availability of infrastructure, international policies and gender issues play an important role in food security at both global and national levels.

2.1.3.1.1 Availability

Physical access to nutritional food for the common people is known as food availability. "Food for the people" is the main theme for food security which provides nutritional food to those people who are under the constraints of food scarcity, due to extreme geological locality and non-availability of renewable resources and provision for conventional energy supply system. The food shortage of a particular region occurs when people are not in a position to effort energy, water and fertilisers for agricultural practices. Food crisis can be easily understood, but to solve the problem, dedicated involvement of local administrative bodies and cooperation of public is highly necessary. Provision of proper storage system and import of food grains are primary and immediate steps to be taken by the government, in collaboration with international agencies. The past history says that food shortage also occurred despite the availability of abundant food. The great hunger of Ireland (1845–1847) and the famine of Bengal (1944) have been mainly due to British political decision to export locally available grain supplies without compensating imports than to production shortfalls [25, 26].

The other serious problems like natural disasters such as drought, flood or fungal infection and political distresses like civil conflict and mismanagement of economic policies discourage the village farmers for production of essential foods. The uneven pattern of food distribution in an area is mainly due to mismanagement of food supply chain by politically influenced bureaucrats.

The environmental, socio-economic condition and administrative operation protocols are interlinked and can be integrated with food crisis problem at regional, national and global levels. Even though irregular supply of fertiliser and lack of proper irrigation may cause food production, the main causes of disruption of fertiliser supply chain and irrigation systems at the time of food scarcity lead to hunger and malnutrition. It has been noted that access to food is more influenced by international market and police trends even though there have been lots of improvement in logistic system network [27]. The framework on market supply chain can be extended to remote areas where the small group of farmers and working community can avail the opportunity for marketing in a wide range of neighbouring places. New agriculture technologies have been the key factors for enhancing food production to keep pace with demand and production under food scarcity problems [28]. This can be well extended to remote villages to support their energy dependence on renewable resources.

Proper planning of water resource conservation and use is primarily an unavoidable factor for food security. However, conflicts over usage of water resources for the cultivation of fuel crops vs. cultivation of cereal and other food crops and use of drinking water vs. industrial use may be threats for long-term food security. The Stockholm International Water Institute (SIWI) emphasises the growing importance of green water, i.e. the water hidden in the ground as soil moisture (while blue water refers to water available in lakes, rivers and aquifers) [29]. With proper amendment to soil structure, irrigation system can be modified for water-use efficiency through Variety of cultivation methods. With the availability of modern technologies and infrastructure development process, water harvesting and recycling of waste water and soil-water management in rain fed systems the resilience of agricultural management system can be strengthened for better livelihood. For this purpose, cooperation of regional water user groups may be helpful in planning and management to reduce the conflicts on water use [30, 31].

2.1.3.1.2 Access

In spite of sincere efforts of many international bodies like the World Health Organization (WHO), United Nations Environment Programme (UNEP), World Food Security and Food and Agriculture Organization (FAO) of the United Nations, hunger and access to food and nutrition are still not satisfactory in numerous countries and regions around the world. Variety of factors like earning of money, the racial or cultural background of certain population, and availability of food for local people are some of the important factors access to food security. Access is ensured when a poverty-stricken region or nation has sufficient resources to produce quality food in adequate amount. This could only be possible when there is availability of adequate amount of quality food, even in adverse climatic conditions. Besides this, the physical, social and environment status are also determinate factors, which catalyse the accessibility. Any drastic changes in these dimensions may cause alarming conditions in productivity and food accessibility. For example, under severe droughts or flood or any other natural calamities, developing countries may be affected. In 2011, there was severe famine in the two regions of Africa. The famines were due to droughts and conflicts that were exacerbated by increased food prices. Over 2 million of the 3.7 million people in need of food cannot be accessed due to regional conflicts. This unusual food crisis was resolve with the aids from the United States and several international voluntary organisations. But to serve those famine-affected people was tough when aid workers have continued to encounter violence in their efforts to deliver resources to those in need.

Disruption in harvest process and increase in price of scantily available food affect the availability and accessibility of food for livelihood. To prevent such incidental misfortune, development of different technical preventive majors is available. However, due to poor logistic facilities in adjusting rural area of an urbanised township suffer a lot to access food distribution, at the time of emergency. The construction of infrastructures such as small dams and reservoirs or water-spreading weirs to hold back water and raise shallow groundwater tables may be immensely helpful to meet any emergency food crisis [31].

2.1.3.1.3 Use and Utilisation

Utilisation is often interpreted as nutrition which is ultimately related to food storage, food processing and food supply chain to reach each and everybody, timely.

Due to the non-availability of quality food, timely, almost a third of children in the developing world are underweight or have stunted growth. Poor nutrition quality has caused about 60% child death worldwide. It has been estimated that the overall economic involvements on undernutrition are about $20–30 billion per year [32], caused primarily by lowered productivity and cognitive capacity and increased healthcare costs. Economic and agricultural productivity are not only the solutions for resolving malnutrition. They can be helpful in caloric intake than dietary diversity [33, 34].

The practice for value chain approach to food utilisation can be adapted by integrating nutrition into value chain selection, planning for well-designed nutrition diet, diversifying dietary habits, promoting production of nutrition-rich food and vegetable other than regular cereal crop cultivation, using nutrition-sensitive value chain analysis, amending dietary style (rich in antioxidant vitamins) instead of taking high-calorie food, following nutrition-sensitive value chain analysis of daily food chart, analysing the implications of project design on nutrition, supporting nutrition-sensitive processing upgradation and identifying opportunities to enhance food safety and value chain efficacy.

Use of food is defined as the potential of physiology and metabolism of body to use nutritional and hygienic food for the growth and development and to keep normal structure and function, with the variation of age and other surrounding physical factors. Use of food mainly depends on the socio-economic aspects of household food and nutrition security and food habits of a particular locality. On the basis of availability of nutritious food in the market, the household has to decide what food to purchase and how to incorporate it in a regular food diet. Another aspect of food use is on the basis of biological activities. This mainly depends on the potential of the human body to take food and convert it for body building. Acquiring of energy is very important to meet the proper calorie for keeping the body fit for day-to-day works. Besides taking nutritional diet, safe drinking water is essential to maintain normal health condition. The groundwater is often contaminated with industrial wastewater and agricultural wastewater, and as a result 884 million people worldwide have no access to adequate drinking water [30, 31, 35].

2.1.3.1.4 Stability

Since 2017, worldwide increase in food insecurity has given big challenge to SDGs for completely eradicating hunger from the global market by 2030. Rapid increases in global instability, conflict and migration as a result of inadequate global food supplies and water scarcity have been catalysing the hunger and poverty problem in increasing magnitude. Presently, the world population is about 7.6 billion and is expected to reach 8.6 billion by the end of 2030 and 11.2 billion by 2100 [36].

Besides this, rising incomes in many low- and middle-income countries are further increasing the demand for food to satisfy the desire for higher-quality, more nutritious and diverse diets. Stability of food security could be possible when food supply chain on the household level remains constant during the year and in the long term. It is important to minimise external risks such as natural disaster and climate change, price volatility and conflicts due to political instability. In order to overcome such problems, it is the moral duty of the concerned government to provide financial aid in terms of crop insurance, e.g. against drought and crop failure as well as the protection of the environment and the sustainable use of natural resources like land, soil and water [37–39].

2.1.3.2 International Organisation for Food Security

Celebration on October 16 for World Food Day is to revive the status of food and nutrition security, at global front, especially in rural areas. This campaign is to catalyse the movement for zero hunger drive and reach the target by 2030. Currently, food insecurity is a nontraditional security threat for poverty-prevailing villages. In reality, the drive for zero hunger is intimidating. The Food and Agriculture Organization (FAO), in 2016, has reported that about 11% of the global population are under the extreme constraint of chronic undernourishment. This case is noticed to be maximum in Asia (FAO, 2017). The FAO data survey report says that undernourishment in South Asia was 9.4% in 2015. This was increased to 11.5% in the year 2016. Wasting and stunting are the common visible symptoms for malnutrition. According to the joint report of UNICEF, the World Health Organization and the World Bank (2018), wasting prevalence in 2017 was estimated about 8% across the globe. The occurrence of malnutrition was surprisingly higher in magnitude in the mountain region. For example, in Hindu Kush Himalaya range (total length is about 2300 km), about 31% of the total population are under the constraints of malnutrition problem. On the basis of WHO report in 2016, about 800 million people who are hungry every day and about 156 million children under the age group of 5 years are suffering from malnutrition [40].

Adverse factors like unusual climatic changes, poverty, extreme geographical locality, scarcity of water resources and socio-economic conditions are mainly responsible for food crisis. In addition, increased inclination towards cash crops and changing food habits restrict the dietary intake of people and may result in malnutrition and even micronutrient deficiencies. The unusual changes in climatic condition catalyse the problem of food insecurity and hunger, both of which have continued to increase in 2017 in some areas of the world.

According to WHO report (2016), the global volume of food wastage is estimated at 1.6 billion tonnes of "primary product equivalents", and out of which, the amount of edible part content in this food wastage is 1.3 billion tonnes [41]. This food wastage is equivalent to 3.3 billion tonnes of CO_2-equivalent GHG released into the atmosphere per year [42].

Further, food wastage is also an emerging challenge that undermines the efforts to end hunger and malnutrition. So, ending food wastage is therefore crucial to achieving zero hunger by 2030.

Under the heading "State of Food Security and Nutrition in the World 2017", the United Nations has released a report stating a five percentage increase in global hunger from 2015 to 2016. The report states that the number of chronically undernourished has reached 815 million people or 11% of the global population. This report was prepared in collaboration with the Food and Agriculture Organization (FAO), the International Fund for Agricultural Development (IFAD), the United Nations Children's Fund (UNICEF), the World Food Programme (WFP) and the World Health Organization (WHO) and provides broader view on hunger, malnutrition and public health.

So, food security is a must for planetary health (i.e. the health of human civilisation and the state of the natural system on which it depends). The scarcity of food alters the life in this habitable world physically and socially by the disruption of micro- and macro-ecosystems surrounding a specific region. So the United Nations, under the scheme of Sustainable Development Goals (SDGs), is targeting to eradicate hunger by 2030. The 2030 Agenda comprises 17 new Sustainable Development Goals (SDGs) which will assist the nations suffering from hunger and poverty by framing guidelines and funding.

The concept of the SDGs was initiated in Rio+20 Conference held by the United Nations, in 2012. The main aim was to frame a set of universally applicable goals that cover the three dimensions of sustainable development: environmental, social and economic. The SDGs replaced the Millennium Development Goals (MDGs) which were initiated in September 2000. The MDGs were targeted to prevent and eradicate extreme poverty and hunger and life-alarming disease and provide educational opportunities to all children, all over the world. The new development agenda applies worldwide to develop peaceful and inclusive societies, create job opportunity for all and challenge the unusual climatic changes.

During the Sustainable Development Summit on 25 September 2015, UN member states adopted the 2030 Agenda for SDGs with the 17 goals for ending poverty, fighting inequality and injustice and handling problems related to climatic changes by 2030 (Fig. 2.6). The 17 SDG's goals include (1) No Poverty, (2) Zero Hunger, (3) Good Health and Well-Being, (4) Quality Education, (5) Gender Equality, (6) Clean Water and Sanitation, (7) Affordable and Clean Energy, (8) Decent Work and Economic Growth, (9) Industry, Innovation and Infrastructure, (10) Reduce Inequality, (11) Sustainable Cities and Communities, (12) Responsible Consumption and Production, (13) Climate Action, (14) Life Below Water, (15) Life on Land, (16) Peace and Justice Strong Institutions and (17) Partnerships to Achieve the Goal (Fig. 2.7).

Fig. 2.7 Pictorial presentation of 17 SDGs

In this connection the United Nations Development Programme (UNDP) also aims to eradicate poverty and reduce inequalities through the sustainable development of nations, in more than 170 countries and territories. So, all the 17 SDGs are associated with the existing objectives of UNDP's Strategic Plan including sustainable development, democratic governance and peace building and climate and disaster resilience. Goals Number 1 on poverty, Number 10 on equality and Number 16 on governance are the primary targets for UNDP's work plan. UNDP mainly supports three aspects common to both UNDP and SDF programmes. (i) UNDP provides support to governments to highlight the new global agenda in national development plans and policies, (ii) supports countries to accelerate progress on SDG targets and (iii) makes the UN's policy expertise on sustainable development and governance available to governments at all stages of implementation. From overall aspect UNDP is deeply involved in all processes where SDG is involved.

Among the 17 SDG goals, the number 2 is mainly targeted to eradicate hunger, to achieve food security, to improve nutrition and to promote sustainable agricultural practice, worldwide. The main salience features include:

(a) To end hunger by the end of 2030, targeting the people, including infants, who have been suffering from hunger in vulnerable situations, and to make access to food and nutrition available for them, around the years.
(b) To end all forms of malnutrition, in the form of stunting and wasting in children under 5 years of age, to provide nutritional needs to adolescent girls, pregnant and lactating women and older person.
(c) To increase agriculture production to double quantity of small-scale food producers, particularly women, indigenous peoples, family farmers and pastoralists and fishers. In addition maximum stress will be given to provide land to landless farmers and to develop local food supply chain system.

(d) To endure sustainable food production by introducing resilient agricultural practice which could result in productivity and production. This could be possible by developing small water reservoir by collecting rainwater or excess water from rivers during rainy season.
(e) To ensure and maintain genetic diversity of seeds, cultivated plants and farmed and domesticated animals and their related wild species, by the end of 2020. This could be possible by soundly managed and diversified seed and plant bank at the national, regional and international levels.
(f) To establish international cooperation in rural infrastructure development, agricultural research and extension service to rural farmers and provision for availing facilities from gene bank (for improved variety of crop plants and livestock) and adapting new agricultural technologies for preharvest and postharvest processes.
(g) To facilitate the functioning of food commodity markets and warehouse for food storage and in order to help limit extreme food price volatility.

References

1. Information brief on Water and Agriculture in the Green Economy. UNW-DPAC (2011)
2. United Nations World Water Development Report 4 (2012) UNESCO, UN-Water, WWAP March 2012, www.unesco.org/new/en/natural-scirmces
3. World Population Prospects: The 2008 Revision, Highlights. UNDESA (2009). https://www.un.org/esa/populations
4. World Agriculture: Towards 2030/2050 – Interim Report – Prospects for Food, Nutrition, Agriculture and Major Commodity Groups. FAO (2006). www.fao.org/fileadmin/user
5. Water crises are a top global risk (2015) World Economic Forum. 16 January 2015 https://www.ircwash.org/blog
6. Four billion people face water shortages, scientists find. World Economic Forum. 17 February 2016, https://www.weforum.org
7. How do we prevent today's water crisis becoming tomorrow's catastrophe? (2017) World Economic Forum. 23 March 2017, https://weforum.org
8. Global Water Shortage Risk Is Worse Than Scientists Thought. Huffingtonpost.com. (2016) 15 February 2016. https://www.huffpost.com
9. Hoekstra AY, Mekonnen MM (2016). "Four billion people facing severe water scarcity" (PDF). advances.sciencemag. American Association for the Advancement of Science
10. Birol F (2017) World energy outlook special report, International Energy Agency (IEA). https://www.iea.org
11. DREI: Off-Grid Electrification (2018) UNDP Currently, a wider gap is left between energy and demand and access. https://www.undp.org
12. Goal 2: Zero Hunger (2015) United Nations Sustainable Development. https://www.un.org/sustainabledevelopment/hunger/
13. Sam R, Stokes D (2010) Energy security. In: Collins A (ed) Contemporary security studies. Oxford University Press, New York
14. de Oliveira KL (2012) Energia Como Recurso de Poder Na Política Internacional: Geopolitica, Estrategia e o Papel do Centro de Decisao Energetica. PhD Thesis, Universidade Federal do Rio Grande do Sul, Brazil. Available at: http://www.lume.ufrgs.br/handle/10183/76222
15. Yergin D (2000) Ensuring energy security. Foreign Aff 84(4):69–82
16. World Energy Outlook (2011). www.worldenergyoutlook.org

17. World Bank (2015) March 14, 2015 Energy Overview [Online]. Available: http://www.worldbank.org/en/topic/energy/overview#1
18. International Energy Agency (2015) March 14, 2015. Energy Poverty [Online]. Available: http://www.iea.org/topics/energypoverty/
19. World Energy Outlook (2018) International Energy Agency, https://www.iea.org
20. https://www.business-standard.com (2018) India's natural gas production improves 0.6% in November 2018
21. International Energy Agency (2016). http://hdr.undp.org/en/countries/profiles/ETH#
22. http://www.iea.org/stats/WebGraphs/ETHIOPIA4.pdf
23. Energy access-International Energy Agency (2019). https://www.iea.org/energyaccess
24. UNICEF (1998) The state of the world's children, focus on nutrition, (http://www.univef.org/sowc98/silent4htm)
25. Smith B (1991) The impact of structural adjustment policies on agricultural transformation and poverty alleviation in Africa. In: James V (ed) Urban and rural development in third world countries: problems of population in developing nations. McFarland, Jefferson, pp 22–37
26. Woodham-Smith C (1962) The great hunger, Ireland 1845–9. Harper and Row, New York
27. Idriss J, Alamgir M, Panuccio T (1992) The state of world rural poverty: an inquiry into its causes and consequences. New York University Press, New York
28. Chattopadhyay B, Spitz P (1987) Introduction. In: Chattopadhyay B, Spitz P (eds) Food systems and society in Eastern India: selected readings. United Nations Research Institute for Social Development, Geneva, pp 1–5
29. Siwi I, Iucn I (2005) Let it reign: the NewfckLRWater paradigm for global food security, FinalfckLRReport to CSD-13. Stockholm International WaterfckLRInstitute, Stockholm
30. Klennert K. 2009) Achieving food and nutrition security, Feldafingen/InWEnt
31. USAID (1995) Food aid and food security policy paper, (http://transition.usaid.gov/policy/ads/200/foodsec/foodsec.pdf)
32. UNICEF, Progress for Children: A Report Card on Nutrition (2006)
33. The World Bank (2006) Repositioning nutrition as central to development: a strategy for large-scale action 27, https:siteresources.worldbank.org
34. Fan S, Brzeska J The Nexus between agriculture and nutrition: do growth patterns and conditional factors matter? www.ifpri.org
35. IICA, Food Security, http://www.iica.int
36. https://www.un.org/development/.../population/world-population-prospects-2017.html
37. National Food Fortification Alliance, Action Plan, undated, https://www.gainhealth.org
38. Feed the Future, African Alliance for Improved Food Processing, https://parthersships.usaid.gov
39. Case Study in Using the DCED Standard: Tofu Production in Indonesia with the VIP (2011), https://www.enterprise-development.org
40. WHO (2016) Ending malnutrition in all its forms? A decade of opportunity, https://www.who.int
41. UNICEF (1998) The state of the world's children, focus on nutrition, (http://www.unicef.org/sowc98/silent4.htm
42. FAO (2016) News article: food wastage: key fact and figures, www.fao.org/news/story/item/196402/icode

Paradigm for Zero Hunger

3

A society or a country can also be developed under the guidance of international development or global development policies for a better style of living, in a sustainable pattern. For better understanding and convenience, countries can be classified as developed country, developing country and least developed country. Still, there are many different theories proclaiming the real stick yards for "development". Generally, it is speculated that development has been mainly synonymous with economic development. "International development" is different from the simple concept of "development". International development is a process that creates growth, progress, positive change or the addition of physical, economic, environmental, social and demographic challenges. The best example in this connection is the development of the Bretton Woods Institutions, immediately after World War II, which was mainly focused on economic growth, alleviating poverty and improving conditions in previously colonised countries (https:www.ted.com). In 2000, the programme for the "Millennium Development Goals" and in 2015 the "Sustainable Development Goals" are the major gigantic steps for developing the living standard of the people, fighting for poverty and hunger, on global front.

From a historical prospective, the second half of the twentieth century has been known as the "era of development" [1]. The main target of development by international development agencies was to bring amendments in the standard of livelihood of previously colonised country by restructuring the social living standard on the basis of globalisation development prospective by establishing free trade policies between so-called "developed" and "underdeveloped" nations. In addition it was also targeted to change colonised habitats as independent states in order to enjoy freedom from poverty, hunger and social insecurity [2, 3]. In order to accelerate such movement, before 1949, the United States had taken first credit by developing the International Bank for Reconstruction and Development (now part of the World Bank Group) and the International Monetary Fund (IMF) in the year 1944. In addition the launch of Marshall Plan was another significant step in operating projects on international development in order to create political and economic bloc in Europe that was allied to the United States. International development trend had

B. K. Behera et al., *Move Towards Zero Hunger*,
https://doi.org/10.1007/978-981-32-9800-2_3

taken a new progressive direction, when, during the 1970s and early 1980s, the World Bank and IMF adapted neoliberal ideas of economists such as Milton Friedman and Bela Balassa [4]. The functional aspect of international development was geared up when the International Labour Organization (ILO), influenced by Paul Streeten, then by UNICEF [5]. Afterwards UNDP put forward the concept of Human Development by the influential work of Mahbub ul Haq and Amartya Sen on human needs and capabilities [6–9].

During the 1990s, some intellectual groups started thinking that the international development theory had reached stalemate [10, 11]; on the other hand, some academics were in impression of international development concept and that function was in the right direction [11] (Fig. 3.1).

By this time the cold war ended, and capitalism had become the dominant mode of social organisation. The US statistics reported that the standard of living on global front was improved, since the end of World War II [12, 13], although a large portion of the global population were still under extreme poverty and fighting for getting relief from hunger and malnutrition.

3.1 International Development

The past two decades have witnessed the various novel ideas (theories) on international development, in order to overcome the chronic problem of poverty, malnutrition and hunger, mostly being suffered by the rural people of developing countries or people living in least developed countries. The conceptual development of such theories is based on a variety of events associated with social inclusion or social

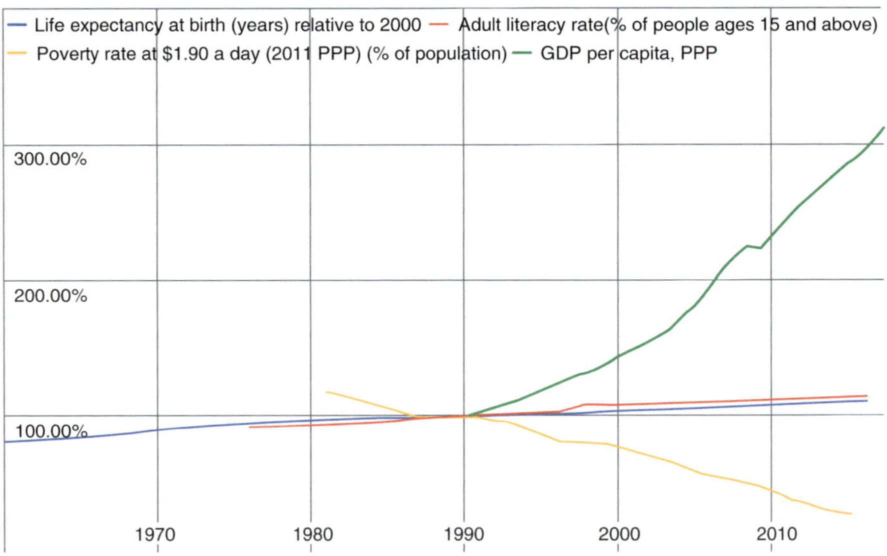

Fig. 3.1 World development indicators improved relatively to the year 1990. (Source Wikipedia)

exclusion, traditional livelihood of rural community and a variety of social scientific disciplines. These theories include:

- Modernisation theory
- Dependency theory
- World-systems theory
- Neoliberalism
- Good governance
- Capability approach

3.1.1 Modernisation Theory

Modernisation of a country or society is the process of upgradation of the standard of living by the application of various modern engineering technologies in a "pre-modern" or "traditional" society to bring changes in their internal structure and function in order to provide a progressive and sustainable pattern of life. The conceptual development on modernisation of a society or country was initially proposed by German sociologist Max Weber (1864–1920), who provided the basis for the modernisation paradigm development by Harvard sociologist Talcott Parsons (1902–1979). In late 1960, the acceptance of modernisation theory gets reduced and was entitled as controversial model [14–26].

3.1.2 Dependency Theory

Dependency theory is a social theory which is based on the economic development of a developing country or a least developing country with the support of a developed country. This theory was first proposed in the late 1950s by the Argentine economist and statesman Raul Prebisch. This theory gained popularity in the 1960s and 1970s. The dependency theory proclaims that the developed country exploits the underdeveloped country for its own prosperity in wealth development and modernisation of standard of living. For example, the underdeveloping country provides cheap labours (both unskilled and highly skilled labours) and raw materials on the world market. These resources are ultimately transferred to developed countries, which are transferred to developing countries as finish goods, at high prices. The result is a vicious cycle that perpetuates the division of the world economy between a rich core and a poor periphery. Fernando Henrique Cardoso, the past president of Brazil (1995–2003), had the opinion of supporting dependency theory, which may bring some level of social structure improvement in a developing nation. But Andre Gunder Frank, German American economic historian, had the opinion that the dependency was the creation of a noncapitalist national economy. Plenty of past literature explain on the beneficial and drawback of dependency theory [27–32].

3.1.3 World-Systems Theory

World-systems theory is mainly based on the conceptual development of the habitable world with its social changes in different countries, mainly focusing on macroscale approach to world history in relation to changes in social structure and function [33, 34]. It mainly refers to the interregional and transnational classification of labour on which the world is classified into core countries, semi-peripheral countries and peripheral countries [34].

Core nations are the most powerful having rich in wealth and independent of outside nations. The core countries are also well to do in all respects with strong bureaucracies system having direct or indirect control over power, strong military forces and well-developed economies. Due to possessing over-smart brain power, they could be able to be highly progressive in engineering technology having strong influence over the less developed noncore nations [35, 36].

The semi-peripheral nations are in a developing process and do not have much hold on international trade with moderate economy condition but are better than the peripheral countries [37]. However, they are in a process of development to compete with core nations. Generally, they prefer to export to peripheral nation and import from core nations [38].

Peripheral nations are least economically developed and poor standards of literacy. These nations mainly supply unskilled labours to the core countries. The nature social inclusion is extremely poor due to the inefficacy government system which is unable to control country's economy and highly influenced by core countries.

The significance of world-systems theory is the concept of humankind evolvement due to the dynamic nature of social, economic and political factors over the time, which, in turn, is responsible for the changes in the overall picture of the world economics.

3.1.4 Neoliberalism

Neoliberalism is police models which interlink politics, social studies and economics for the improvement of economic structure and function of private sector by transferring control of economic factor from public sector. It believes in free-market capitalism and away from government spending, regulation and public ownership. According to neoliberalism, there should not be any legal restriction on capitalism. In the nineteenth century, this form of economic liberalism is used to be known as laissez-faire. The main target of this policy is economic liberalisation such as privatisation, austerity, deregulation, free trade and reduction in government expense by giving maximum scope to private sector [39–41].

3.1.5 Good Governance

Good governance means the process of decision-making and its implementation. The term governance is applicable to different actors like corporate governance, international governance, national governance and local governance. The main actors involved in decision-making may be a government of a nation or state. In rural areas the actors are influential landlords, associations of peasant farmers, cooperatives, NGOs, research institute, religious leaders, finance institutions, political parties, the military, etc. who are involved in decision-making or play a critical role in decision framing. All actors, except the government and military, are part of "civil society". Good governance should have eight main characteristics, which include the following: it should be participatory, consensus-oriented, accountable, transparent, responsive, effective and efficient and equitable and inclusive and follow the rule of law.

In general, good governance is supposed to be an ideal model for the benefit of a nation, but in practice very few countries and societies have achieved the goal of good governance in its totality. However, to develop sustainability in social inclusion and standard of living and eradicate hunger and malnutrition, practicable action should be taken by various actors who are in the helm of affairs in decision-making for the success of SDG-like programme.

3.1.6 Capability Approach

The capability approach is an economic theory developed during the 1980s. It is supposed to be an alternate approach to welfare economics [42]. The two core claims of capability approach are the following: (i) freedom to achieve well-being of an individual is primarily morally important and (ii) freedom to achieve well-being is to be understood in terms of people's capabilities. It is economist-philosopher Amartya Sen and philosopher Martha Nussbaum pioneered the conceptual development of capability approach. The capability approach claims that freedom to achieve well-being is the matter of capability of an individual and the type of life he or she able to lead. The capability approach is flexible in its scope of application having multipurpose framework, rather than a precise theory of well-being [43–46].

3.2 Global Goals for Eradication of Hunger and Poverty

3.2.1 Millennium Development Goals (MDGs)

In 2000, the millennium declaration was held at the UN headquarters, New York, where 189 UN member states participated and made the decision to eradicate poverty and hunger from the global map. This major decision was famous as "Millennium Development Goals (MDGs)" having targeted to eight social issues which include

Fig. 3.2 A pictorial presentation of Millennium Development Goals

(1) to eliminate extreme poverty and hunger; (2) to achieve global primary education; (3) to empower women and bring gender equality; (4) to reduce child mortality; (5) to promote maternal health; (6) to fight malaria, HIV/AIDS and other diseases; (7) to promote environmental sustainability; and (8) to develop a universal partnership for development (Fig. 3.2).

The implementation of these eight agenda of MDGs was decided to be initiated from 2001, and with a gap of 5 years, an assessment summit should meet to have glimpse on overall progress of MDGs, on world front. All the eight agenda of MDGs are interlinked with each other. To have a healthy society is the topmost priority of MDGs, as better health peruses the literary movement among children which is ultimately responsible for empowerment of youth. Gender equality is also essential to acquire better health. Poverty and hunger eradication and environmental conservation could only be possible with better community health programme and its successful implementation. The tentative periods of 15 years, from the date of announcement of MDGs (2000), were estimated to achieve success in this mega event for poverty, hunger-free and healthy society, globally. Thus, United Nations Secretary-General Ban Ki-moon has proclaimed as “a change to change the course of history”. All the eight goals as mentioned in MDGs are universal in nature to be achieved by the global community, and it should be the moral obligation of each and every beneficiary state to extend the same service to those who have not. As such they may represent an extension of the concept of human right. The applicable significance of MDGs is measurable in its activities in terms of social indicators based on statistical data collected and maintained by respective organisations in each relevant field, mostly by UN agencies like the Organisation for Economic Co-operation and Development (OECD), International Monetary Fund (IMF) and World Bank. Most of the developing countries try to implement MDGs in order to amend the standard of social livelihood which should be free from poverty and hunger. Science 2000 with the implementation of MDGs, Africa, as a whole could be able to upgrade social standard of living. Although considerable achievements have made on many of the MDG targets universally, progress has not been uniform across developing regions and nations, leaving substantial gaps. For example, sub-Saharan Africa was left behind with least progress in MDGs. Still, millions of people are under the constraint of poverty and hunger due to their age, sex, ethnicity, disability and geographic locations.

The implementation of MDGs helps in orientation of new development era with the tie-up of developed and developing countries. Chronic issues like poverty, hunger, malnutrition and some environmental problems persisting in developing world started reducing in its intensity with the strong commitment of developed countries for supporting timely financial aids and other necessary assistance in order to upgrade the social standard of living and social inclusion process. The basic philosophy for implementing MDGs included (i) quantification of requirement to be undertaken by MDGs programme in developing countries, (ii) assessing whether the existing financial aids and other resource will be adequate for developing countries and (iii) possibility of enhancing financial grand on the requirement basis of developing countries. On the basis of the first report, it was understood that progress in MDGs project was satisfactory in East Asia since 1990 (the baseline year for MDG targets), although the goals were targeted to globalise the MDGs programme. It was observed that the least developing countries like sub-Saharan Africa are still in the grip of poverty and hunger. The first part of MDGs programme was completed without any much hope. Thus, in 2005, at Gleneagles it was decided to increase the financial aid to African countries, having the hope for progress by the end of 2010. But, due to some more localised crises, and global recession of 2009, it was not possible to achieve the MDGs target. Under such constraint, and little time to go until 2015 MDG target date, efforts shifted towards the poorest and more fragile countries lagging much behind in leading normal life without hunger and poverty. UNDP played a constant and active role throughout these days by campaigning UN MDGs programme to give global shape to this programme.

3.2.2 Sustainable Development Goals

In September 2015, the world leaders at historic UN Summit had taken decision for universal welfare by passing 17 Sustainable Development Goals (SDGs) having the target to erase poverty, hunger and malnutrition from global map and to bring strong social inclusion by establishing gender equality, education for all, right for minimum facility to maintain standard health and provision of clean water like 17 social welfare agenda to be implemented on global front (Fig. 3.3).

The framing of all these universal welfare activities is based on the success of the MDGs. The significant aspects of these new goals are that they call for action by all countries, poor, rich and middle-income to promote prosperity while protecting the planet. The basic motto of MDGs is both the poverty eradication and zero hunger are complement to economic growth and addresses a range of social needs including education, health, social inclusion and youth empowerment. In addition tackling of climate change and environment protection are most unavoidable issues while going

Fig. 3.3 A schematic presentation of Sustainable Development Goals (SDGs)

ahead with MDGs programme. UNDP played a constant and active role throughout these stages by campaigning and contributed through the Inter-Agency Expert Group (on MDG indicators) to global monitoring programme. Since 2008, UNDP and Department of Economic and Social Affairs (DESA) Jointly involved in reporting on MDG 8, (MDG Gap Task Force). Lateran, due to the sincere involvement of UNDP, was elevated to the United Nations Development Group (UNDG) and subsequently to the Chief Executive's Board, under the direct supervision of the Secretary-General and in close collaboration with the World Bank. The SDGs programme has wider scope, going beyond the "social" goals as projected through MDGs programme. The SDGs cover all aspects of social developmental parameters like economic, social inclusion and environmental sustainability (Fig. 3.3). The SDGs not only are more ambitious agenda, to resolve the problem of poverty or hunger, but also have a wide range of targets like healthcare, education, environmental sustainability and other aspects. The most salient feature of this agenda is universal applicability for development cooperation and peace.

In order to enhance the efficacy of SDGs, a common organising framework was formed in the name of Mainstreaming, Acceleration and Policy Support (MAPS). The intention of framing such working group is to generate awareness among all relevant actors and help governments land the agenda at national and local level; ultimately to mainstream the agenda into their national plans, strategies and budgets; and to bring awareness among national stakeholders which include government departments, civil society, parliamentarians, the media and businessmen. In addition, information should be provided on the working policy of the United Nations Development Assistance Framework (UNDAEF) in the plan of implementation at national level.

It has been realised that global package of services can be better way operated with little amendment in working policy at national level. For example, UNDP will provide strong and sincere support for SDGs 1, 10 and 16 (poverty, inequality and governance) on need base of a nation, mostly selection of goals related to the environment, climate change and women's and girls' empowerment. The overall action planning of SDGs can cover three interlinked thematic areas: (i) development planning and industrial sustainable growth, (ii) economic recovery and environment and (iii) natural capital. For development planning and to increase functional efficacy, UNDP supports in identifying, accessing, combining and sequencing finance for sustainable development. This is because development planning and inclusive sustainable growth are keys to eradicate poverty, to enhance social inclusion, to remove health disparities and to achieve sustainable development.

For economic recovery and environment protection, UNDP works globally for the disaster-affected society and community in order to relieve and protect situations for short-term economic and long-term economic development and livelihood recovery. The main theme of this programme is towards reducing vulnerability and building resilience. The three critical steps for economic recovery plan include (i) to bring livelihood stabilisation to normal condition with the simultaneously action plan for relief, recovery and development; (ii) to bring economic stability and employment opportunity and (iii) to bring sustainability in economic and environment during unavoidable shocks due to unusual climatic changes and natural disaster.

Natural capital accounting or environmental-economic accounting is a means of application to bring realisation among the actors of both the public sector and private sector in order to gain an understanding of the interaction between the economy and environment. By means of this tool, we can measure the state of ecosystems, flows of ecosystem services as well as changes in stocks and flows of natural resources in relation to economic changes. The natural capital accounts (NCAs) consist of supply and use table, functional accounts (e.g. environmental protection expenditure accounts) and asset accounts for natural resources. This report explains overall aspects of potential uses of NCAs in national policy process for the SDGs.

Unusual climatic changes, social exclusion, clash at national and community levels, food crisis, inadequate energy availability, persistence pockets of poverty and hunger and rapid urbanisation are some of the major challenges to achieve SDGs, according to the recent UN report declared in New York in 2018. On the basis of this report, it has been understood that conflict and climatic change were major factors responsible for poverty and hunger in many of the extreme geographic locations. The intensity of hunger has been increased to 815 million in 2016 than the earlier report, 777 million in 2015. Thus, there are now approximately 38 million more hungry people in the world. In 2017, the world has witnessed the costliest North Atlantic hurricane season on record, driving the global economic losses attributed to the disasters to over $300 billion. However, it has been noticed that in some patches of world, the standard of living in rural areas has upgraded, from earlier condition. One of the most significant results with the application of SDGs is the reduction in mortality rate by almost 50% in most of the least developing countries. The population with access to electricity has more than double between 2000 and 2016.

But, it has been observed that in 2015, 2.3 billion people were lacking of proper sanitation service. The malaria case has been increased to 892 million as compared to data available in 2013, which was about to 210 million.

Still, the habitable world moves in progress to achieve the SDGs, in confrontation with various issues social and climatic challenges, with the passing of time.

3.3 Global Food System

Food system consists of the governance and economics of food production, its sustainability, infrastructure involved in feeding a population, food processing (growing, harvesting, packing, transporting marketing), minimisation of food waste and interlink of food procurement and environment management (Fig. 3.4).

In addition it includes how the food affects health and well-being, including nutrition, obesity and food safety. Food system is not the synonym of food security and nutrition value improvement. Food system has a significant value in achieving social, economic and environmental goals as well. Presently, the world food productivity is 17% more than 30 years ago. Still, it is a wonder that close to billion people go to sleep hungry in every night. This is mainly due to unequal distribution of agricultural land or lesser income to purchase adequate food. In addition, lacking of food storage warehouse where the rural farmer can store excess amount of food, at the time of food scarcity. Thus, food system is a wider field of applications which cover: nutrition, food quality, health, community economic development and agriculture. The entire food system is operated within and influenced by social, political, economic and environmental aspects. In addition the management and food system operation need human resources that provide labour, researchers and technical manpower [47–49].

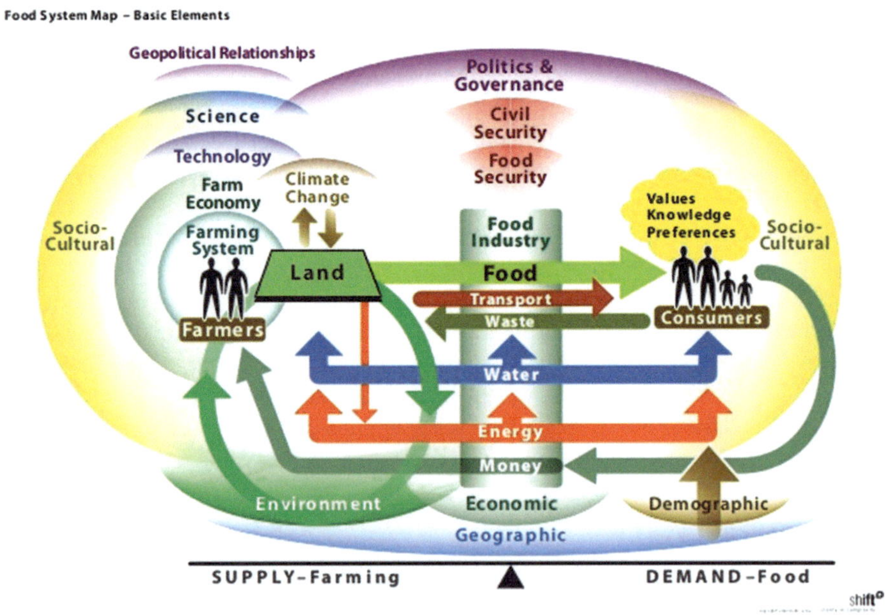

Fig. 3.4 Schematic presentation of global food system

3.3.1 Conventional Food System

Conventional food system is based on the utilisation of vertical integration, economic specialisation and world trade. The efficacy of this system is mainly dependent on maximum order placement to minimise the cost at consumer level and enhance overall production. In addition, low cost of fossil fuel, manufacturing of low-cost fertiliser, food processing and supply chain management are important aspects of food system development. In practice, conventional food system food production is more than the alternative system. Excess use of fertiliser, nonpoint sources of pollution and greenhouse gas emission are major factors for bringing adverse impact on conventional food system [50]. The globalisation of food production associated with the conventional system, often responsible for loosing traditional food system [51].

3.3.1.1 Conceptual Development of Food System

Practice of agriculture and production of surpluses food for future need were started with the development of habitable world. At later stages, these surpluses of food became the basis factor for developing ancient civilisation, particularly those in the Fertile Crescent. The Fertile Crescent is a crescent-shaped region located in the Middle East, being surrounded by modern-day Iraq, Israel, Palestinian Territories, Syria, Lebanon, Egypt and Jordan as well as the south-eastern fringe of Turkey and the western fringes of Iran. This region is known as "cradle civilisation", the place where the inhabitant begins the cultivation of crops by clearing the natural vegetation. System of trading in exchange of food stuff emerged in East Asia, North America, Southern America and sub-Saharan Africa [52, 53].

The history says how new food system was introduced during the conquest of Alexander the Great, the Crusades, the expansion of Islam, the journeys of Marco Polo and the exploration and colonisation of the Americas by Europeans. Modern practice on industrialised agriculture was resulted after World War II and was ultimately responsible for development of models on food production, food supply chain and conventional food system [54].

Urbanisation has been a major catalytic factor in changing food system. Urbanisation is resulted in bringing together large populations within a specific locality and needs efficient governance to monitor in delivering nutrient-rich food for the public. With the advance of time, new tools and technique are developed for timely delivering food with perfect packing system for safety and security of food stuff. During the Iron Age and the Roman Empire, the Iron Age and the Roman Empire brought expanding empires and the beginning of global food systems, including regional specialisation in products traded throughout empires. Food systems began to be organised on a grand scale to feed larger cities and fuel local economies. Trade networks for grain, nuts, oils, fruit and wine developed using both road systems and sailing routes. Standardised weights and measures were established along with the expansion of money and accounting.

3.3.1.2 Conventional Farming

Food system is extremely dynamic in nature. Since the beginning of agricultural practice, food systems have been frequently evolving with new appearance and challenges. During Stone Age (around 2.5 million years ago), people used to protect their body from the cool weather through animal skin. For this purpose most have learned handling weapons for hunting animals in order to get animal skin as warm cloth. During the end of Stone Age, people began to farm and domesticate animals (around 2000 BC). In due time, the Stone Age people started realising the need of better tools and techniques to improve farming practice. So, man switched to bronze technology which was supposed to begin in 3200 BC in Southern Europe (Greece) and in 32,000 BC in Central Europe. Copper would have also been used for improving farming technology during this period as well. The then civilisation was not satisfied with bronze or copper for farming purposes. Iron was chosen as a better option due to its abundant nature, throughout the world. Thus, iron began to replace bronze around the world beginning roughly in 1000 BC, which was known as the starting era of Iron Age. Meanwhile, the ancient civilisation must have developed innovative idea on better way of farming practice by developing new tools with the use of iron. The first systematic farming was started 5000 years ago by the Celts civilisation which is presently known as England. The Celts used two different forms of farming. One practice was arable farming which is a process of ploughing for seed plantation. The arable farming was mainly practised for growing oats, barley, millet, rye and corn, although wheat was the most popular in England during the Iron Age. Later in the Iron Age, Celts began to clear forests to make use of more land as their population increased, and farmlands were in great demand. Surplus food grains were stored in granaries (food warehouse) and used as and when needed. Celts also used fertiliser for better productivity and to lower the price. Celts also used pastoral farming where larger tracts of land were used for the raising of animals for meat, milk and clothing. The animals Celts raised during the Iron Age were early derivations of more commonly known domesticated animals. The Soay sheep was the English breed of the Iron Age.

Conventional agricultural methods have only been in development since the late nineteenth century and did not widely spread until post-World War II. Conventional agriculture is a reliable resource which is energy intensive, but highly productive. In conventional agriculture practice, chemical fertilisers, pesticides and herbicides are used to increase productivity. In addition use of genetically modified crops/organisms, concentrated animal feeding operation, heavy irrigation, intensive tillage and concentrated monoculture production are also encouraged. Conventional farming is usually contrasted to organic farming, otherwise known as sustainable agriculture or permaculture. Organic farming is a practice based on integrating cultural, biological and mechanical practices that promote cyclising of resources, maintain ecological niche and conserve biodiversity.

In the 1970s, most of the developing countries intensified in conventional agricultural practice, which resulted in negative environmental impact. In order to overcome such problem, the concept of sustainable agricultural practice was encouraged in rural areas adjunct to urbanised locations. Awareness was developed for organic farming (use of minimum synthetic fertilisers), international environmental movement and maintaining ecological balance for a sustainable ecosystem.

The main target of conventional food system is to achieve lower food cost with greater food variety. By reducing the production expenses, those savings can be then passed onto the consumer by means of lowering the selling price of food stuff. The advent of industrial agriculture and availability of infrastructure have catalysed the practice of conventional food system. However, conventional food systems are largely based on the availability of inexpensive fossil fuel, which is necessary for mechanised agriculture, the manufacture or collection of chemical fertilisers, processing of food products and the packaging of the foods. So, with the passing of time, agricultural practice and domestication of animals set the stage for permanent settlement in a specific geographical location instead of moving around unlike earlier hunter-gatherers. Different grain cultivation emerged at different geographic location, adapted with the regional climatic condition, for example, maize in Mexico, rice in China and wheat and barley in the Middle East.

3.3.1.3 Community Food System

A "community food system" is defined as a sustainable food production system consisting of production, harvesting, postharvesting, distribution and consumption in an integrated framework in order to enhance the environmental, economic, social and nutritional health of a specific location. The global communities are associated with diversified projects that develop connections between different parts of the food system. The main theme of these projects is to involve local people in multiple food system issues such as (i) to make availability an adequate quality food for community people, (ii) to bring sustainability in family farming practices by using minimum chemical fertiliser, (iii) to develop food supply chain system for easy availability of food stuff in local market, (iv) to improve the lifestyle of labourers and farmers and (v) to frame better food and agriculture policies that could promote local food production, processing and consumption. It means community food system offers farmers an opportunity to acquaint with direct marketing, local processing and other value-added activities, consumers to develop opportunities to reconnect with their food supply and communities to gain opportunities to strengthen their social and economic health by creating meaningful jobs and recirculation of social and financial capital locality.

The community food system projects are mainly based on collaboration efforts from various sectors of a community which may be formal or informal in nature. The participation of a broad cross-section of the community is essential for the project to be representative and contribute to the growth of a community food system. Besides this, the involvement of local stakeholders in a community food system project to understand in depth the local issues related to food production, utilization and marketing is highly necessary. The stakeholders in mutual understanding with local people should develop a coalition. Coalition means "individuals or organisations working together in a common effort for a common purpose to make more effective and efficient use of resources" [55].

Overall food production resulted from the joint effort of farmers, fishers, hired workers and residents residing in a community. On location base, a community occupation involves crop plating, harvesting of raw food materials, seafood collection and processing, packing of food stuff and marketing. The increase in output can attribute mostly to an increase in the quality of labour, capital and technology input. During the mid-20th century it has been noticed that inputs used to produce growth in food output, dramatically increased. Due to development of new tools and techniques in the food cultivation, harvesting and postharvesting process, a significant reduction in labour input has been noticed. In addition, the use of capital inputs has remained roughly the same, and the use of purchased variable inputs has more than doubled. During 1970, the application of capital equipment started growing rapidly, while the land inputs slowly declined. The use of chemical fertilisers started showing impact on higher productivity. In the mid-1970s, the use of fertiliser accounts for a significant portion of the increased use of purchase inputs, and this trend was continued till 2011. It appears that the reduced impact in productivity growth from a decline in the use of labour and land inputs has been offset by the positive impact of increased use of other inputs (e.g. technology, computerisation, fertiliser, pesticide).

3.3.2 Alternative Food System

Alternative food system is outside the scope of conventional agriculture. It is having dynamic structures of food production, distribution, acquisition and consumption that are generally in opposition to the conventional food system.

3.3.2.1 Local Food System

Local food is a movement of people who prefer to consume food for normal livelihood available in and around the location of a community. Local food movements are mainly based on local food producer and food consumers of a specific geographic location in order to develop more self-sufficient and self-resilient food networks. This practice helps in improving local micro-economies, health management, environmental conservation and development of social inclusion [56]. Besides this, it helps in development of food supply chain within community and also neighbourhood [57, 58].

Local food is a movement of people who prefer to consume food for normal livelihood, available in and around the location of a community. They are different from the industrial food systems by operating with reduced food transportation and more direct marketing. These practices help in developing personal relation-based local marketing systems where people get the opportunity for face-to-face interaction and develop trust in trading process [59]. Besides this, a cordial relationship between consumers and farmers develops, which may lead to improvement in productivity and reduction in retail price [60, 61].

The best examples of local food systems include community-based agriculture, farmer markets and farm to school programmes. Farm to school is a programme in the United States through which schools procure locally produced farm-fresh foods rich in nutrients such as fruits and vegetables, eggs, honey, meat and beans on their menus.

3.3.2.2 Organic Food System

Organic food systems are based on the minimum or no use of chemical fertilisers made with synthetic ingredients or sewage sludge, bioengineering or ionising radiation. In addition, the livestock is reared without the use of antibiotics or growth hormones. Organic meat, poultry and dairy products come from animals that are provided with food without antibiotic or growth hormones. In marketing organic food, transparency on food information is vital for organic food. Although organic foods are less risk than the conventional foods, still there are no common international standard rules and regulations to monitor organic food from the site of production to retail market.

The standard of organic food varies worldwide, but the salient features of organic farming are common worldwide and include cycle of resources, promote ecological balance and conserve biodiversity. In the beginning of the twenty-first century, the European Union, the United States, Canada, Mexico, Japan and many other developed countries require producers to obtain special certification to market their food as organic. Selling food with organic brand is regulated by governmental food safety authorities, such as the National Organic Program of the US Department of Agriculture (USDA) or European Commission (EC).

To certify an organic product, each nation should have an accredited agent to inspect organic food of each producer or manufacturer for compliance with that market's standard. For example, the US Department of Agriculture (USDA) has framed national standards that "organic" food must meet, whether it is grown in the United States or imported from abroad. Meanwhile, organic market has been picking up continuously. In 2002, Japan became the first to accept organic products certified under the USDA standard. Recently, the United States recognised the ability of Canada, New Zealand, Denmark and the United Kingdom to accredit agents who will certify organics under the USDA standard.

3.3.3 Cooperatives in Food System

Success in food system could only be possible with the cooperation attitude and efforts between cultivators and consumers. In farming cooperative farmers pool resources, to either cultivate their crops or market them. Besides this, consumer cooperatives also include the process of running cooperative grocery stores and maintaining food warehouse for surplus food with no profit or loss basis. Recently, community-supported agriculture has been picking up rapidly. By such practice community members can buy a share in a farm's harvest and extend their cooperation as farm labour for both consumers' and producers' end of food systems. In cooperative food system, the benefit and risk are mutually shared by the members to bring sustainability in food system management.

In a cooperative food system, the flow of capital expenditure is mainly controlled through the producer, harvester, consumers and multi-stakeholder. The overall food systems like farming, fishing, buying, value-added, purchasing, food hubs, distribution, marketing, food service management, equipment, labour and land access are managed by the members, belonging to different fields of specialisation.

Community's effort for food system development is not a new process. In the past few decades, cooperatives have played a central role in US agriculture since the Grange movement for promoting cooperatives among farmers. In the early twentieth century, US Congress supported the right of farmers and fishermen to form cooperatives to market, process and distribute products. Since the 1970s consumer cooperatives encouraged for natural, organic and local foods.

Producer cooperatives have been in practice in American agriculture for the last 150 years. In a producer cooperative, the owners of the cooperative are farmer and fishermen at individual level, and manage to provide service such as marketing, aggregation, distribution and value-added processing. Producer cooperatives help the farmers and fishermen in terms of equipment, land, cold storage or labour on no loss and no profit basis.

Farmer cooperatives also play an important role in food system development and management. Food quality is a social construct and farmer cooperatives' key player in quality management. In order to bring efficacy, farmer's cooperatives should be linked with other network actors to develop strong marketing coverage and also to bring amendments on the basis of consumer's requirement. But, developing alliances with other network actors is difficult because of cooperative's limited communication capacity and lack of economic and social capital. In order to bring improvement at farmer's cooperatives, it is necessary to develop at governance level strong policies that support farmer cooperative capacity building at the micro level which could able to improve the performance of smallholder farmers and their cooperatives in the expanding domestic quality food market.

3.3.4 Fair Trade

Fair trade is an organisation which supports the producers in better trading at both national and international fronts. Members of fair trade movement plead for higher prices to exporters accompanied by improving social and environmental standard. The basic functional aspect of fair trade is the trading of commodities or products from developing countries to developed countries, besides the consumption in the domestic markets (e.g. Brazil, India, Bangladesh). Mostly the fair trade goods are handicrafts, coffee, cocoa, wine sugar, fresh fruits, chocolate, flowers and gold [62, 63].

The fair trade believes in transparency in dealing and promotes greater equality in international trading partnerships through amicable dialogue and respect. It also gives trace on sustainable development by mutual understanding on trading terms and conditions. The practice of fair trade is mainly motivated in three commandments, which include the following: (i) there should be cordial relation between consumers and producers, (ii) the current global trade practices exist to promote the unequal distribution of wealth between nations and (iii) buying products from producers in developing countries at a fair price is a more efficient way of promoting sustainable development than rational charity and aid.

Generally, fair trade movement follows informal association of four international fair trade networks which include the Fairtrade Labelling Organization International,

World Fair Trade Organization (WFTO), Network of European Worldshops and European Fair Trade Association (EFTA).

In order to have quality control trading, there are several recognised fair trade certifiers which include Fairtrade International (formerly called FLO, Fairtrade Labelling Organizations International), IMO, Make Trade Fair and Eco-Social.

3.4 Food System Supply Chain Management

Food supply chain management is a process of distribution of food stuff from production site to retail dealer. This is a process in which products move from the farmer level to the consumers to meet consumer's requirements, reliably in terms of quality, quantity and fair price. Thus the entire process of food supply chain management involves the movement of local foods from the farm to the consumer, including marketing, markets, distribution pattern, aggregation, processing, packaging, preparation, resource recovery and food waste disposal (Fig. 3.5).

The commercial significance of supply chain management (SCM) is to bring competitiveness and customer's satisfaction. It is an integral part of a firm's success.

3.4.1 Food Supply Chain in Rural Areas

Mainly, farmers from the rural part of developing countries, peripheral region of developing countries and least developing countries have to confront the practice of food supply chain due to lack of proper logistic. Still, a great majority of food consumed is procured from rural-urban, rural-rural and urban-rural supply chains. Any disruption of food supply chain creates great concern over food security system management. In general, the unusual climatic changes are mainly responsible for the disruption of food supply chain. The second major factor for hunger is lack of logistic system to meet the rural food demand at the time of serious climatic change or natural calamity. The Food and Agriculture Organization of the United Nations reported that one-third of all food produced worldwide, amounting about 1.3 billion tons, is lost or wasted, although an astonishing number of people are suffering from hunger (Fig. 3.6).

The food wastes in industrialised countries amounts US$ 680 billion, and in developing countries this amount goes up to US$ 310. Among the food wastes, fruits, vegetables, plus roots and tubers are contributing as major parts of food wastes. Generally, worldwide, roughly 30% cereals; 40–50% root crops, fruits and vegetables; 20% oil seeds, meat and dairy plus; and 30% fish are dumped as food wastes. The main reason for such big loss in food stuff is lack of the provision of proper logistic system linked to farms and markets. In addition, lack of food warehouses in rural areas and least developing countries causes damage to food stuff. In spite of development of modern agriculture engineering and technology, about 40% of losses happen at postharvest and processing levels. In developing countries food losses and waste occur mainly at earlier stages of food value chain processing.

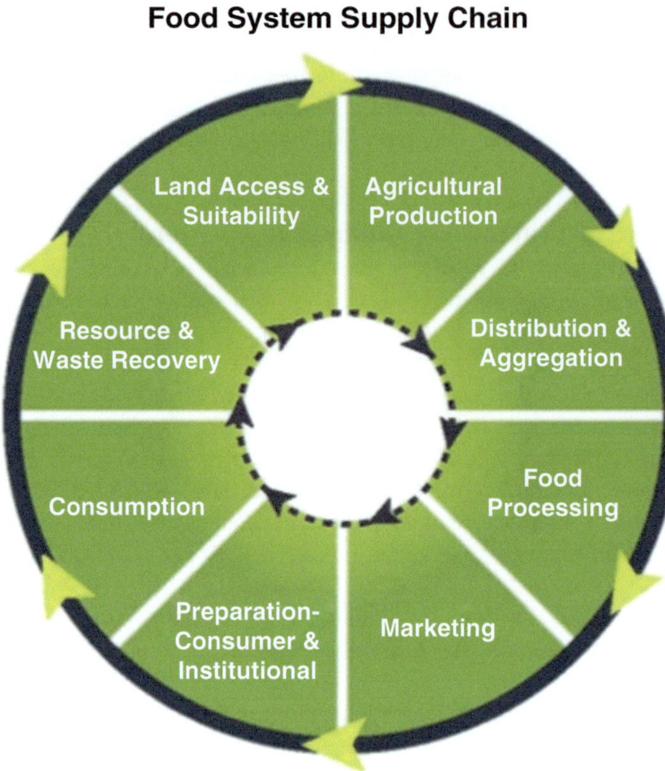

Fig. 3.5 Food system and supply chain management process

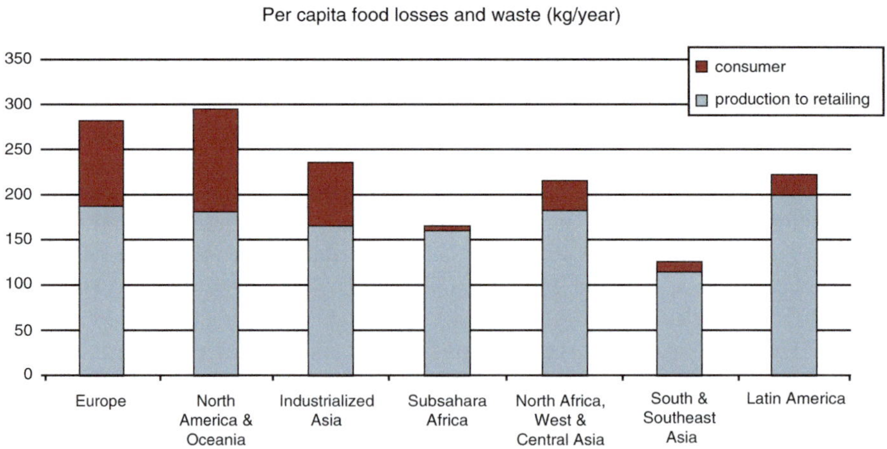

Fig. 3.6 Worldwide food wastes (Source: The Food and Agriculture Organization of the United Nations)

However, in medium- and high-income countries, food is wasted and lost during supply chain. Thus, farmer-buyer cordial relation can be helpful in order to overcome such problems. Food loss or waste is unintentional decay or damage of food that does not reach its final point of consumption, timely. An example would be a truckload of tomatoes spoiling in heat shock on the way from production site to market. The same way chicken that is processed, transported to the grocery store, if not sold timely, is treated as waste. Food waste or food loss means an extravagance of water, soil and labour and overall management expense responsible for food production. So, food that is grown but not consumed also impacts worldwide prices. The developing countries pay less attention for saving surplus food and use the same in food crisis. For example, cereal grains are more expensive in developing nations where people are least able to afford a price increase. So, UN's Food and Agriculture Organization has noticed that limiting food waste globally could reduce the need to raise more food by 60%. In other words, the issues on hunger can be dramatically controlled with good governance by providing reasonable financial assistance to farmers for developing food warehouses and developing the provision of logistic system for timely movement of food from farms to markets.

3.4.2 Disruption of Food Supply Chain

The food supply chain bridging between farm production site to marketing sites (local, neighbouring rural area and urban markets) and related organisations. It is a cluster of three segments which includes:

(a) Farm production (from upstream process to downstream process) mainly consists of (i) farm input or upstream process (cultivation, harvesting, postharvesting and packing), (ii) farm segments (mainly associated with farm input process [64]), (iii) the midstream process (wholesale, logistic system for transport) and (iv) downstream process consisting of retailing (stores and communities)
(b) Food supply chain improvement (mainly with R&D of institutions) and organisations of each segment
(c) Integration of supply chain input with fertiliser distribution and fuel to transports

The three-dimensional cluster of food chain is very much susceptible to short-term climatic shock and also long-term climatic changes [65].

The rural livelihoods mainly depend on the supply inputs to farmers and sales channels to urban areas. The off-farm components like transport, commerce and processing and handling are key sources of employment in rural areas for a majority of households and sources of capital income for farm investments. Transformation of traditional food supply chain in developing regions of rural areas has been in progress, since the past several decades. The transformation has involved moving from traditional short fragmented chains to transitional and modern forms in developing regions.

In urbanised areas, supply chain is widely covered and linking rural areas. The structure and function of supply chain in urbanised area are picking up modernised infrastructure like supermarkets having the provision of wide variety of items like processed food, and non-grain products like milk, meat, fish, fruits and vegetables with hygienic packing and processing [66, 67]. But, the critical point is that short-term climatic shocks and long-term change could be responsible for disruption of the complex cluster of supply chain. So it is necessary to counteract with the unusual changes due to short-term shocks and long-term climatic changes. In addition good governance should have a well-planned precaution to overcome such problem of both developing regions and rural areas.

3.4.2.1 Sustainability

The FAO definition of sustainable agriculture development is "the management and conservation of the natural resource base, and the orientation of technological and institutional change in such a manner as to ensure the attainment and continued satisfaction of human needs for present and future generation. Such development conserves land, water, plant and animal genetic resources, is environmentally non-degrading, technically appropriate, economically viable and socially acceptable".

Food system and supply chain are mainly dependent on the limited natural resources. The availability of food stuff and their quality are mainly dependent on good governance whose innovative policy decision may be helpful in eradicating food crisis at the time of short-term climatic shocks or long-term changes in climate. In order to overcome such problem, agriculture sectors must be improved with suitable amendments that can confront climatic change to ensure food security for all. The sustainability has three dimensions: economic, environmental and social. On the basis of these three aspects, sustainability in farming system brings together the farm workers and neighbourhoods on a single platform with mutual benefits. The SDG 2 is framed to end hunger by achieving food security with nutritional diet and promote sustainable agriculture.

Application of modern agriculture technology has certainly increased the productivity in specific regions of developing countries. Still, many parts of peripheral regions of developing countries and least developing countries have been under the cloud of hunger and poverty. But, scarcity of water to bring sustainability in agriculture is a big challenge. Water scarcity is a big problem for every continent. About one-fifth of world population spend their livelihood in physically deficit of water. About 1.6 billion people (one quarter of world population) are suffering from the inadequate infrastructure development for supply water from river and aquifer. The demand of water is in increasing magnitude and growing at more than twice the rate of population increase in the last century, although there is no global water scarcity as such. Water scarcity is both a natural and a human-made phenomenon [68, 69].

A geographical locality having less than 1.000 m^3 per person confront water scarcity, and below 500 cubic metres are known as absolute scarcity (Fig. 3.7).

About 700 million people in 43 countries are under water scarcity. By 2025 this figure will reach 1.8 billion. The present dynamic nature in climatic changes may lead to water scarcity problem for half of the world population, if proper majors will

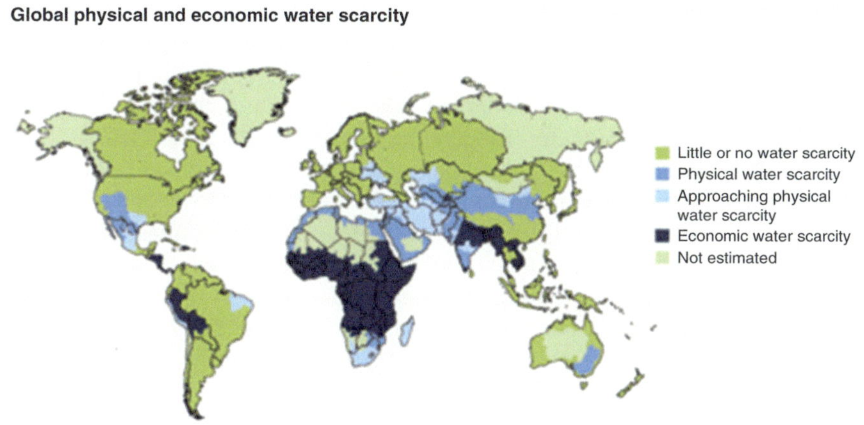

Fig. 3.7 World water development report 43 (Source: World Water Assessment Programme (WWAP), March 2012)

not be taken, timely. The worst to suffer for water scarcity will be African regions. In addition, water scarcity in some arid and semiarid places will displace between 24 million and 700 million people. Sub-Saharan Africa has the largest number of water-stressed countries of any region. Pressure on water resources is increasing in several parts of the world, especially in China, India, Pakistan, the Middle East and many countries and regions of Africa.

So, the major factors of water scarcity include pollution and overuse of water, conflicts in river water distribution, drought and inefficiency of government in controlling and distributing water for drinking and agricultural use. Water scarcity causes hunger, diseases and other health complications, poor sanitation, lack of education and, ultimately, poverty. But, the problem of water scarcity can be eradicated with the collective efforts of each and every one of us playing our own responsibility for resolving this global crisis. Following are few major steps to be taken for minimising water scarcity problem.

3.4.2.1.1 Infrastructure Development

Water conservation, restoration and supply infrastructures are key major factors for resolving water crisis. For these purposes, developing water supply line for pure drinking water, spreading of canal system all around the water resources, storing of excess water by constructing dames, developing rainwater harvesting systems in rural areas and recycling of industrial wastewater for irrigation are important projects to be systematically managed by good governance with the implementation of policies linked to water conservation and utilisation.

Besides this, monitoring of natural infrastructure like landscape management for water conservation and restoration, flood control and aquifer storage and recharge and providing clean and abundant supply of water are also unavoidable requirement for minimising water crisis.

3.4.2.1.2 Groundwater Recharge

Groundwater is a potential source of water, if it is handled properly. Artificial groundwater recharging is a pretty old technique, which has been in use since few decades. The process of water recharge could have been possible by means of human-induced process or by natural process through hydrological cycle. Groundwater recharge is carried out directly via injection wells or spreading basins or as a result of human activities including waste disposal and irrigation (Fig. 3.8).

Aquifer recovery and storage is also a better option in overcoming the problem of water scarcity. The surface water can be infiltrated into unconfined, shallow aquifer. Presently, this process is amended to direct injection into deeper aquifer systems, including those having poor-quality groundwater, for the storage of freshwater. The main problems with such practice are clogging of pores from particulates, chemical precipitation and biofilm formation.

3.4.2.1.3 Rainwater Harvesting

Rainwater is a natural source of water which can be used during water crisis, by proper storing. This can be practised via the use of a water catchment and collection basin. Rainwater collected directly can be used for drinking and other household works. In many parts of Africa, water catchment and rainwater harvesting have proven to be important sources of water during water crisis (Fig. 3.9).

The main benefit of rainwater harvesting practice is the storage of safe and clean drinking water in rural areas where provision of community water filter system is not available. This makes it possible for rainwater to be harvested in large quantities.

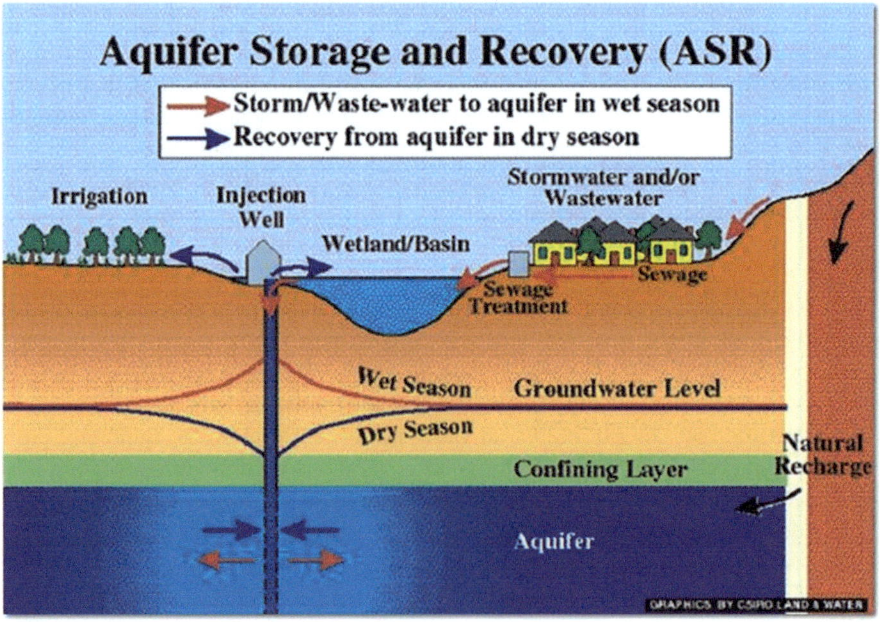

Fig. 3.8 Schematic diagram of aquifer storage and recovery

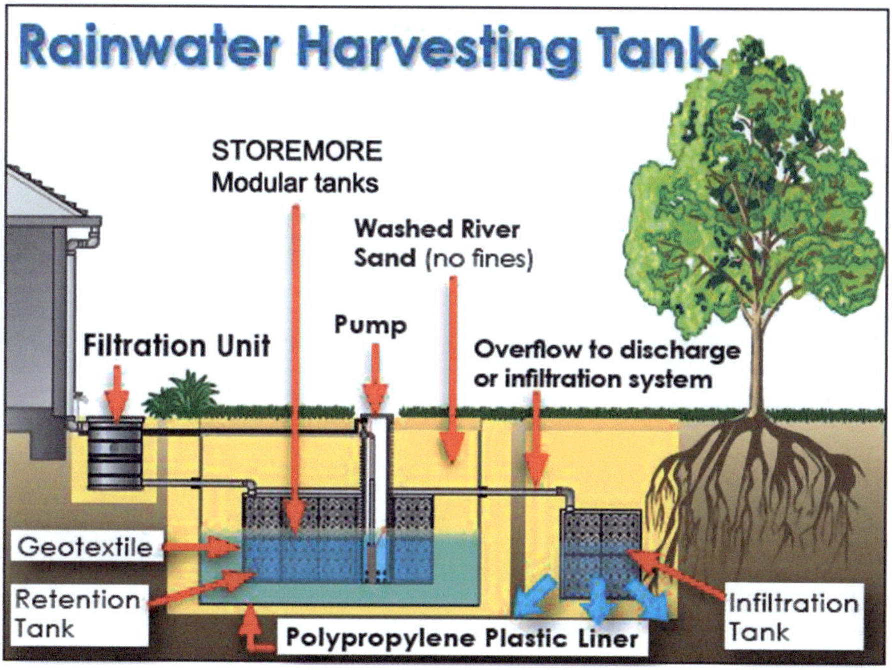

Fig. 3.9 Graphic presentation of rainwater harvesting process

Currently, three different water harvesting systems are in practice: (i) the rooftop catchment, (ii) the surface catchment and (iii) the runoff farming.

3.4.2.1.4 Rooftop Catchment

This system is consisting of iron-sheet spreading over the rooftop with the provision of gutters and downpipes linked to storage system, mostly fibrocement tanks.

3.4.2.1.5 Surface Catchment System

Surface water harvesting includes all systems that collect and conserve surface run-off after a rainstorm or in intermittent streams, rivers or wetlands for storage in open ponds and reservoirs. This can supply water for direct household use, irrigation, livestock and aquaculture. Compared to rooftop catchment technique, ground catchment technique is a better option for collecting more quantity of water from a large surface area and can meet water demand during dry period. Due to infiltration into the ground, various techniques are adapted to increase the runoff within ground catchment areas: (i) clearing or altering vegetation cover, (ii) increasing land slope by means of artificial ground cover and (iii) reducing the soil permeability by soil compaction and application of chemicals.

Fig. 3.10 Indian village women moving far away from their village to fetch drinking water

3.4.2.1.6 Runoff Farming

Runoff farming is simply the use of water harvesting for agriculture purpose. The harvested runoff water from uncropped areas is directed to a cropped area; this technique is known as runoff farming. Mainly, factors like depth of soil profile, depth of plant roots, texture, structure, infiltration rate and water holding capacity of soil directly or indirectly impact runoff farming.

The runoff water farming would be beneficial to millions of women in developing and least developing countries, if the efficacy of runoff farming is exploited in a proper way. By adapting this practice, the women would be saved from the burden of trekking too far away for water (Fig. 3.10).

In arid region of Africa, runoff water farming saves the rural people during dry period when water scarcity used to be at peak. Runoff water collected from the rooftop can be stored for drinking without filtration. Runoff farming complements this harvesting of drinking water, boosting crop yields by catching runoff on fields, improving direct infiltration and preventing soil loss. The practice of water harvesting has a long history, but in recent year, it has drawn significant attention to solve water crisis at different geographical locations.

3.4.2.2 Ensuring Food Security

In spite of so much efforts by international organisations like MSGs and SDGs for eradication of poverty, malnutrition and hunger, the number of chronically undernourished people in the world is estimated to have increased from 777 million in 2015 to 815 million in 2016 (FAO, AFAD, UNICEF, WFP and WHO 2017). Whatever progress made since the last decade is noticed to be uneven. The short-term climatic shock and long-term climatic changes have been causing adverse impact on agriculture, at global level. So, it is necessary to understand the different aspects of food insecurity before planning to bring new amendments in food security policy matters, at global level. FAO has defined four dimensions of food security: food availability, economic and physical access to food, food utilisation and stability over time. Detail on food security has already been described in the second chapter of this text.

3.4.2.3 Impact of Climatic Change on Agriculture

The impact of climatic changes on agriculture depends on the location of a specific region and surrounding environment. For example, climate is expected to increase temperature and precipitation variability, reduce the predictability of seasonal weather patterns and increase the frequency and intensity of severe weather events, such as flood, cyclones and hurricanes. The Intergovernmental Panel on Climate Change (IPCC 2014) pointed out that changes in climate and carbon dioxide concentrations will enhance the distribution and increase the competitiveness of important invasive weeds.

Climatic change may render the cultivated area unsuitable for crop production, and some tropical grassland may become more arid. For example, 10–20 million hectares of land in the region of sub-Saharan Africa alone was adversely effected by climatic changes, and the crop production gets reduced from double cropping system to single cropping [70, 71].

Besides this, climatic changes may bring directly adverse impact on agriculture production, which may ultimately be responsible for the downfall of economic and food security. The major negative aspects of climatic changes are yield reduction, increased yield variability, displacement of crops and livestock breeds and loss of agricultural biodiversity and ecosystem stability. Ultimately, climate changes may increase the gap between developed and developing countries.

3.4.2.4 Agriculture's Impact on Climatic Change

Mainly, agriculture sectors are major contributor for greenhouse gas emission. Agriculture, Forestry and Other Land Use (AFOLU) contributes major greenhouse gas.

A decade back the greenhouse gas emission from AFOLU was accounted for about 22% of total emission [72] (Fig. 3.11).

These emissions are due to natural process and agriculture practices. So, it is necessary to think about how and to what extent agriculture production system and food systems can contribute to climate change mitigation without compromising food and nutrition security. This could be possible by decoupling production growth from emissions growth. This involves reducing emissions per kilogramme of food output. Crop and livestock production and forest sectors can act as "sinks" for capturing and storing carbon. For this purpose sustainability in crop production and forestry development is an alternate option.

It has been observed that intensive agriculture practice (mainly monocultures) for feeding farm animals is ultimately responsible for emitting the highest amount

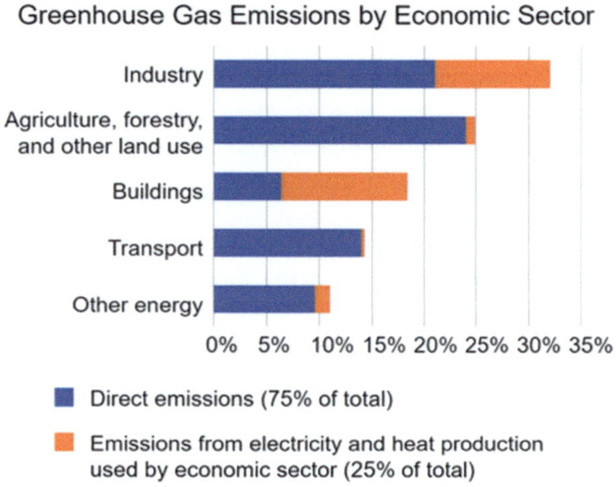

Fig. 3.11 Human greenhouse gas emissions by sector, in the year 2010. "AFOLU" stands for "agriculture, forestry and other land use"

of carbon dioxide, the main greenhouse gas. This quantity can be compared only to the sum total of the carbon dioxide emitted by all forms of transport.

The global meat production has tripled over the last four decades with an increase of 20% in the last decade. The meat consumption of developed countries is about double than the meat consumption of developing countries. Large-scale meat production also causes global climatic changes. Animal waste releases methane and nitrous oxide, the greenhouse gases that are 25 and 300 times more potent than carbon dioxide, respectively. More than 95% of the soy produced in the world is consumed by farm animal (mostly bovines). According to a study conducted by the Chalmers University of Technology in Goteborg, 1 kg of bovine meat requires 200 kg of CO_2 emission (https://www.lifegate.com).

The overall release of greenhouse gases (CO2, CH_3 and N_2O) from agriculture sectors is most significant [73–75]. The microbial decay and soil organic matter burning of plant litter contributes high amount of greenhouse gases [76, 77]. Methane is also produced by the decomposition of organic materials with oxygen-deprived condition, stored manures and rice grown under flooded conditions [78]. Agriculture greenhouse gas (GHG) fluxes are complex and heterogeneous, but the active management of agricultural systems offers possibilities for mitigation.

Agriculture mitigation activities are mainly dependent on social, economic and environmental issues, rather than by means of any high technical methods. Other mitigation methods carry more uncertainty impact on sustainable development. In agriculture sectors, mitigation and adaption complement each other, depending on geographical location. The last few decades have witnessed the agricultural mitigations for reducing greenhouse emission. This also includes non-climatic polices like macroeconomic, agriculture and environmental policies [79].

Organic agriculture plays an important role in mitigating climatic changes. By promoting aerobic microorganisms and high biological activity in the soils, the climatic mitigation can be minimised. The oxidation of methane can be increased due to activity of aerobic microorganisms. Changes in ruminant diet can also reduce methane production. However, investigation on minimising biological methane production by aerobic microbes in paddy field is still in its infancy. The capacity of organic agriculture to contribute the mitigation of climate change can be considered as a supplementary benefit to its primary goal of sustainable land use. This can be achieved by increasing soil productivity, conservative food security and biodiversity [80].

References

1. Thomas A (1990) Poverty and the end of development in Allen, Thomas (2000) Browne
2. The International Development Research Centre (2009). www.idrc.ca
3. Ziai A (2017) Post-development 25 years after the development dictionary. Third World Q 38(12):2547–2558
4. Todaro MP, Smith SC (2009) Economic development, 10th edn. Addison-Wesley, Essex. isbn:978-1-4058-7424-3
5. Cornia GA, Jolly R, Stewart F (1988) Adjustment with a human face, vol I. Oxford University Press
6. Sen A (1977) On the approach to planning against hunger. FAO Review on Agriculture and Development, Ceres. 58 (July-August)
7. Sen A (1989) Co-operation, inequality and the family. In: Population and Devlopment review, development and population: institutions and policy, vol 15. Published by Population Council, pp 61–76. https://doi.org/10.2307/2807922
8. Sen A (1988) Sri Lanka's achievements: how and when. In: Bardhan P, Srinivasan TN (eds) Rural poverty in South Asia. Oxford University Press, Delhi
9. Sen A (1987) Hunger and entitlement. World Institute of Development Economics Research, Helsinki
10. Schuurman FJ (2000) Paradigms lost, paradigms regained? Development studies in the twenty-first century. Third World Q 21(1):7–20
11. Escobar AE (1995) Encountering development: the making and unmaking of the third world, Edition: STU-Student edition, Princeton University Press, Princeton, Page 312
12. Fukuyama F (1989) The end of history? Natl Interest Cent Natl Interest 16:3–18
13. Doney and Wroe (2005) Trade matters. DFID, ISBN-10:1861927371, ISBN-13: 978-1861927378
14. Bernstein H (1971) Modernization theory and the sociological study of development. J Dev Stud 7(2):141–160
15. Black C (1966) The dynamics of modernization: a study in comparative history
16. Brown RD (1972) Modernization and the modern personality in early America, 1600–1865: a sketch of a synthesis. J Interdiscip Hist 2:201–228
17. Chin CC (2011) Modernity and national identity in the United States and East Asia, 1895–1919 (Kent State University Press; 160 pages; An intellectual history of American, Chinese, and Japanese views of modernity)
18. Dixon SM (1999) The modernisation of Russia, 1676–1825. Cambridge University Press, Cambridge, UK. isbn:978-0-521-37961-8
19. Gavrov S, Klyukanov I (2015) Modernization, sociological theories of. In: James D (ed) Wright (editor-in-chief), international encyclopedia of the social & behavioral sciences, vol 15, 2nd edn. Elsevier, Oxford, pp 707–713. isbn:978-0-080-97086-8
20. Gilman N (2004) Mandarins of the future: modernization theory in Cold War America. Johns Hopkins University Press
21. Gilman N (2003) Mandarins of the future: modernization theory in Cold War America (2003). 329 pp
22. Goorha P (2010). Modernization theory In: the Oxford Research Encyclopedia of International Studies, Oxford University Press

23. Shipping H, Zhong Y (2006). Political civilization and modernization in China: the political context of China's transformation
24. Inglehart R, Christian W (2005) Modernization, cultural change and democracy: the human development sequence. Cambridge University Press, New York. isbn:978-0-521-84695-0
25. Misa TJ (2004) Brey P and Feenberg A (eds.) Modernity and technology
26. Tipps DC (1973) Modernization theory and the comparative study of societies: a critical perspective. Comp Stud Soc Hist 15:199–226
27. Paul J (1997) Post-dependency: the third world in an era of globalism and late capitalism. Altern Soc Transform Hum Gov 22(2):205–226
28. Shandra JM, London B, Williamson JB (2003) Environmental degradation, environmental sustainability, and over urbanization in the developing world: a quantitative, cross-national analysis. Sociol Perspect 46(3):309–329
29. Arno T (2003) Social cohesion, sustainable development and Turkey's accession to the European Union: implications from a global model. Altern: Turk J Int Relat 2(1):1–41
30. YSo A (1990) Social change and development: modernization, dependency, and world-systems theory. SAGE Publications, Newbury Park, pp 169–199
31. Torkil L (2018) The global perspective: reflections on imperialism and resistance. Kersplebedeb, Montreal, p 321
32. Michelle W (2014) The end of the developmental state? Routledge, p 44
33. Wallerstein I (2004) World-systems analysis. In: Modelski G (ed) World system history, Encyclopedia of life support systems (EOLSS), developed under the auspices of the UNESCO. Eolss Publishers, Oxford
34. Barfield T (1997) The dictionary of anthropology, Wiley-Blackwell, 1997, ISBN 1-57718-057-7, Google Print, p.498–499
35. Lechner F (2001) Globalization theories: world-system theory
36. Wallerstein IM (2004) World-systems analysis: an introduction. Duke University Press, Durham, pp 23–24
37. Immanuel W (1974) The modern world-system in: capitalist agriculture and the origins of the European world-economy in the sixteenth century. Academic, New York
38. Daniel C (1977) Social change in the twentieth century. Harcourt Brace Jovanovich, New York
39. Smith N (ed) (2019) Neoliberalism. Encyclopædia Britannica
40. Goldstein N (2011) Globalization and free trade. Infobase Publishing, USAID, USA, p 30. ISBN 978-0-8160-8365-7
41. Springer S, Kean B, MacLeavy J (eds) (2016) The handbook of neoliberalism. Routledge, London, p 2. isbn:978-1-138-84400-1
42. Boas TC, Gans-Morse J (2009) Neoliberalism: from new liberal philosophy to anti-liberal slogan. Stud Comp Int Dev 44(2):137–161
43. Sen A (1985) Commodities and capabilities. North-Holland Sole distributors for the U.S.A. and Canada, Elsevier Science Pub, Amsterdam
44. Robeyns I (2005) The capability approach: a theoretical survey. J Hum Dev 6(1):93–117
45. Sen A (1992) Inequality re-examined. Clarendon Press, Oxford
46. Sen A (2009) The idea of justice. Allen Lane, London. pp. xxviii + 468
47. Qizilbash M (2008) Amartya Sen's capability view: insightful sketch or distorted picture? In: Comim, Qizilbash and Alkire (eds.), pp. 53–81
48. A Primer on Community Food Systems: Linking Food, Nutrition and Agriculture. Farmland Information Center. www.FoodptintsandFoodsheds.org
49. Ericksen PJ (2008) Conceptualizing food systems for global environmental change research. Glob Environ Chang 18(1):234–245
50. Maxwell S, Slater R (2003) Development policy review. Food Policy Old and New 21(5–6):531–553
51. Schnitkey GD, Miranda M (1993) The impact of pollution controls on livestock crop producer. J Agric Resour Econ 18:25–36
52. Kuhnlein HV, Receveur O (1996) Dietary change and traditional food systems of indigenous peoples. Centre for Nutrition and the Environment of Indigenous Peoples, and School of Dietetics and Human Nutrition, McGill University, Quebec

53. Manning R (2004) Against the grain: how agriculture hijacked civilization. North Point Press, New York
54. Toussaint-Samat M, Bell A (1994) A history of food. Blackwell Publishing, Cambridge, UK
55. Welch R, Graham R (1999) A new paradigm for world agriculture: meeting human needs, productive, sustainable, nutritious. Field Crop Res 60:1–10
56. Clark R (1992) Building coalitions. The Ohio Centre For Action on Coalition Development, Ohio Cooperative Extension Service, The Ohio State University, 1M-95968
57. Feenstra G (2002) Creating space for sustainable food systems: lessons from the field. Agric Hum Values 19(2):99–106
58. Martinez S et al (2010) Economic research report number 97: local food systems concepts, impacts, and issues. Economic Research Service. Wiley online Library
59. Dunne JB, Chambers KJ, Giombolini KJ et al (2011) What does 'local' mean in the grocery store? Multiplicity in food retailers' perspectives on sourcing and marketing local foods. Renew Agric Food Syst. Cambridge University Press 26(1):46–59
60. Marian B (2008) Supermarket chains narrow their sights. The New York Times
61. Prevor J (2010) Jim Prevor's perishable Pundit. http://www.jimprevor.com
62. Mitchell AD, Lewis M (2006) The Lure of the 100-mile diet. Time Magazine
63. Brough D (2008) Briton finds ethical jewellery good as gold, Reuters, NA, http://en.wikipedia.org
64. Moseley WG (2008) Fair trade wine: North Africa's post-apartheid vineyards and the YouTube global economy. Globalizations 5(2):291–304
65. Reardon T (2015) The hidden middle: the quiet revolution in the midstream of Agrifood value chains in developing countries. Oxf Rev Econ Policy 31(1). Spring):45–63
66. Reardon T (2016) Growing food for growing cities: transforming food systems in an urbanizing world. The Chicago Council on Global Affairs, Chicago. April. https:www.thechicagoecouncil.org
67. Tschirley D et al (2015) The rise of a middle class in urban and rural east and southern Africa: implications for food system transformation. J Int Dev 27(5):628–646
68. Swinnen JFM (ed) (2007) Global supply chains, standards, and the poor. CABI Press, Oxford
69. Human Development Report (2006) UNDP hdr.undp.org
70. FAO (2007) Coping with water scarcity. Challenge of the twenty-first century. UN-Water
71. Fischer G et al (2005) Socio-economic and climate change impacts on agriculture: an integrated assessment, 1990–2080. Philos Trans R Soc B-Biol Sci 360(1463):2067–2083
72. Schmidhuber J, Tubiello FN (2007) Global food security under climate change. Proc Natl Acad Sci 104:19703–19708
73. FAO (2011a) Potential effects of climate change on crop pollination, by M. Kjøhl, A. Nielsen, and N.C. Stenseth, Rome
74. Cole CV et al (1997) Global estimates of potential mitigation of greenhouse gas emissions by agriculture. Nutr Cycl Agroecosyst 49:221–228
75. International Panel on Climate Change (IPCC) (2001) Climate change: the scientific basis. Contribution of working group I to the third assessment report of the intergovernmental panel on climate change. Cambridge University Press, Cambridge, UK
76. Paustian K et al (2004) Agricultural mitigation of greenhouse gases: science and policy options. Council on Agricultural Science and Technology (CAST) report, R141 2004, p. 120
77. Janzen HH (2004) Carbon cycling in earth systems – a soil science perspective. Agric Ecosyst Environ 104:399–417
78. Smith P (2004) Engineered biological sinks on land. In: Field CB, Raupach MR (eds) The global carbon cycle. Integrating humans, climate, and the natural world. Island Press, Washington, DC, pp 479–491
79. Mosier AR et al (1998) Mitigating agricultural emissions of methane. Clim Chang 40:39–80
80. Smith P et al (2007) Policy and technological constraints to implementation of greenhouse gas mitigation options in agriculture. Agric Ecosyst Environ 118:6–28

Health for All

4

According to the latest FAO food price index, food prices are back on the rise, causing an increase in global poverty for the first time in nearly two decades (Fig. 4.1).

Poverty, food prices and hunger are, globally, inextricably linked. Poverty ultimately leads to poor health and hunger. Not every poor person is hungry. But, almost all hungry people are poor. So, the first three goals of SDGs focus on issues like no poverty, zero hunger and good health and well-being. So, in order to bring awareness on universal healthcare, the 7th of April of each year marks the celebration of

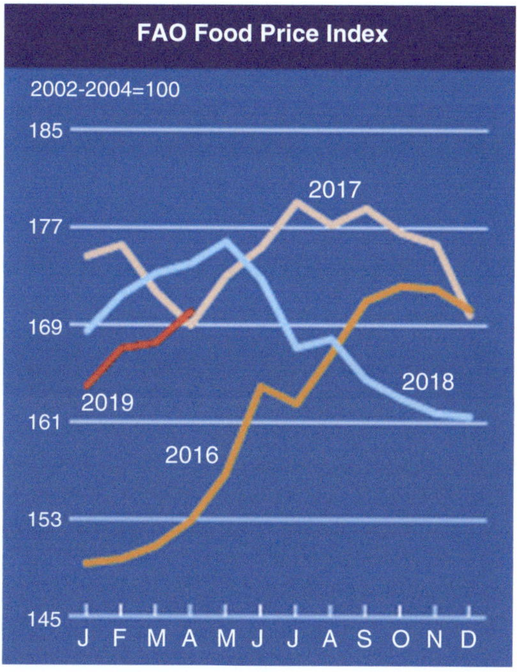

Fig. 4.1 Food price index. (with courtesy from the FAO)

B. K. Behera et al., *Move Towards Zero Hunger*,
https://doi.org/10.1007/978-981-32-9800-2_4

World Health Day. Important health-related issues like mental health, maternal and child care and climate change are taken care on a priority base. The World Health Day celebration, mainly peruse for maintaining tempo which extend beyond the celebration day itself and focus worldwide attention on universal healthcare. This is why, since 2015, worldwide campaign on the most ambitious Sustainable Development Goals (SDGs) has been lunched having specific targets to be implemented in all countries, irrespective of their level of growth and development. In this connection, the World Health Organization (WHO) plays significant role in monitoring global health trends. Since 2016, the World Health Statistics has been highlighting on the updated status of 36 health-related SDG indicators. Although successful implementation of SDGs targets is in progress, still uneven progress in universal healthcare worldwide is a big challenge. The under-five mortality rate has improved, dramatically. But, the recent WHO reports (2016, https://www.wikigender.org) say about 15,000 children die before the age of 5. According to the "World malaria report 2017", unprecedented global success in malaria control was achieved, but there was an estimation of five million more malaria cases in 2026 than in 2015. This could have been due to inadequate funding at both domestic and international levels, resulting in major gaps in coverage of insecticide-treated nets, medicines and other life-saving tools. Although the risk of dying from cardiovascular disease, chronic respiratory disease or cancer has reduced since 2000, an estimated 13 million people under the age of 70 still died due to these diseases in 2016. To gain momentum towards the SDGs is possible if good understanding between government and governance will be maintained in implementing SDGs. In addition, the government has to find out reliable and efficient NGOs or institutes to operate health-related projects being funded by international organisation like the World Health Organization (WHO), the United Nations Children's Fund (UNCF), the United States Agency for International Development, the Centers for Diseases Control and Prevention (CDC), CARE International and Population Services International.

4.1 Conceptual Development of Universal Healthcare

In 1977, the World Health Assembly complies with the theme that the WHO should achieve the goal of universal health consciousness by 2000 to lead a socially and economically productive life which will allow a level of health capable to participate actively in the day-to-day social life of community in which they live. The most significant decision taken by the World Health Assembly is the Global Strategy for "Health for All" by the year 2000. According to the WHO, "'Health for All' does not mean an end to diseases and disability, or that doctors and nurses will care for everyone. It means that resources for health are evenly distributed and that essential health care is accessible to everyone. It means that health begins at home, in schools, and at the workplace, and that people use better approaches for preventing illness and alleviating unavoidable disease and disability. It means that people recognized that ill-health is not inevitable and they can shape their own lives and the lives of their families, free from the avoidable burden of disease".

The programme Health for All gained significant improvement worldwide both in health status and access to healthcare. Majority of the member states (about 147 out of 166) were actively involved for the first evaluation in 1985. In 1997, 158 out of 191 member states showed good performance with respect to health indicators, as approved by the World Health Assembly. Some indicators were more widely covered than others, e.g. 90% of countries reported on immunisation, but only 30% on access to local health service. Overall motivation of the members of state was noticed to be commitmental in achieving the Health for All goals, and most countries have endorsed at the highest level the necessary policies and strategies. In order to increase the efficiency of implementation of the project, the existing healthcare policies are being reoriented to a health system based on primary healthcare, taking into account the role of the individual, the family, the community and the local non-government organisations (NGOs), as well as the health personnel. Maximum attention has been paid to women's health. The World Health Assembly in its first phase (1977–1985) of implementation of "Health for All" programme achieved remarkable success in primary healthcare which includes immunisation against the eight EPI (Expanded Programme on Immunisation) target diseases (diphtheria, whooping cough, tetanus, measles, poliomyelitis and tuberculosis); trained attendance at childbirth; local health services; and clean water supply and excreta disposal facilities. In this connection the gap between developed and developing countries has been significantly reduced, although less satisfactory result was noticed in least developing country. In developing countries 65% of pregnant women have access to antenatal care services and 53% to skilled attendance at delivery. In rural areas, 75% of the population have access to clean drinkable water supply and 34% to adequate sanitation. In 1970, the coverage of immunisation programme was merely 5%. This figure has been significantly improved to 80% in 1996 with an impressive impact on children's health. But these improvements in least developing countries (LDCs) were noticed to be poor. Since 1975, the average life expectancy at birth was shocking (<30%). But, in 1997, 106 countries reported that about 64% of their respective country had an average life expectancy at birth of above 60 years; an infant mortality rate was below 50% per 100% live births and an under-five mortality below 70% per 1000 live birth.

In order to operate successfully the mission for "Health for All", adequate funds for primary health services are unavoidable factors. Countries adapt various ways and means to finance health system, but few countries, even the most affluent, are satisfied with the distribution of financial resources between promotive and therapeutic services. But, still in developing countries and least developing countries, provision of primary education and training on health is extremely poor. So, with proper understanding between government and governance, suitable policies should be undertaken to bring health awareness in public in a community. Due to the complex nature of health systems, expensive maintenance practice, and emerging of new medical technologies allocation of funds becomes more difficult. Reproducibility and comparability of R&D results are basic requirement to success in primary healthcare. Thus, the WHO gives emphasis on quality control and quality assurance for safety and security of individual health in community.

4.1.1 Primary Healthcare (PHC)

Primary healthcare (PHC) is a practice of taking healthcare as a whole of society on need base approach to health of individuals, families and communities. It covers the broader aspects of health which includes physical, mental and social health and well-being.

Primary healthcare is a sustainable practice for need-based health service throughout the lifespan, not for any instantaneous or specific diseases. Primary healthcare guarantees the service for the people with a compressive package having target for promotion and prevention to treatment, rehabilitation soothing care for day-to-day fitting with good health for a normal life.

Article 25 of the Universal Declaration of Human Rights says: "Everyone has right to a standard of living adequate for the health and well being of himself and of his family, including food, clothing, housing and medical care". However, interpretation of primary healthcare is a variable expression but having the basic theme is to be scientifically sound and socially acceptable methods and technology, which can be accessible to all members of families in a community. The PHC is a set of policies framed by an ideal governance giving maximum emphasis on human development, based on the economic, social and political aspects. The World Health Organization (WHO) has viewed PHC under three major issues which include the following.

4.1.1.1 Empowering People and Communities

In 1978, Alma-Ata International Conference on primary healthcare passed a resolution that people should be given priority in developing policies and programmes that affect their health [1]. Later on, this became a core concept of the WHO's goal for *Health for All* [2]. The Alma-Ata Conference campaign for "Primary Health Care movement" in association with professional bodies, institutions, governments and civil society organisations, researchers, and grassroots organisations capable of undertaking politically, socially and economically issues related to primary health policies, mainly of developing countries and least developing countries [3–5].

4.1.1.2 Multisectoral Policy and Actions

Multisectoral actions can be defined as the social and economic factors that influence the health of a population at the local, national and global levels. The main challenges being faced by multisectoral actions are identified as inefficient governance entrenched with political and administrative corruption, widespread clientelism, lack of citizen voice, weal social capital and lack of trust and respect for human rights. However, some World Health Assembly (WHA) and United Nations General Assembly resolutions are acting as external driving forces in favour of intersectoral actions for health.

4.1.1.3 Essential Public Health Function

Public health functions are commonly known as "all public and voluntary entities that contribute to the delivery of essential public health services within a jurisdiction". This implies community health and well-being by the cooperation and help of all entities of a community or state are recognised in assessing the provision of public health services. The public health system includes local- and state-level public health organisation, healthcare provider, public safety and security agencies, charity organisation for human services, nongovernment organisations (NGOs), cultural organisation, philanthropic organisation and environmental agencies (Fig. 4.2).

The following are the ten essential public health services (Fig. 4.3):

Monitor health status to identify community health problem.
Investigation and diagnosis of community health problems.
Introduction of primary healthcare instruction among the public.
Campaign for community partnerships to identify and solve public health problem.
Governance's action plan for supporting individual and community health.
Implementation of law and regulation to protect public health.
Ensure need-based health services well in a timely manner.
Assure competent public and personal healthcare workforce.
Assess efficacy of community healthcare.
Development of R&D for innovative solution of health problems.

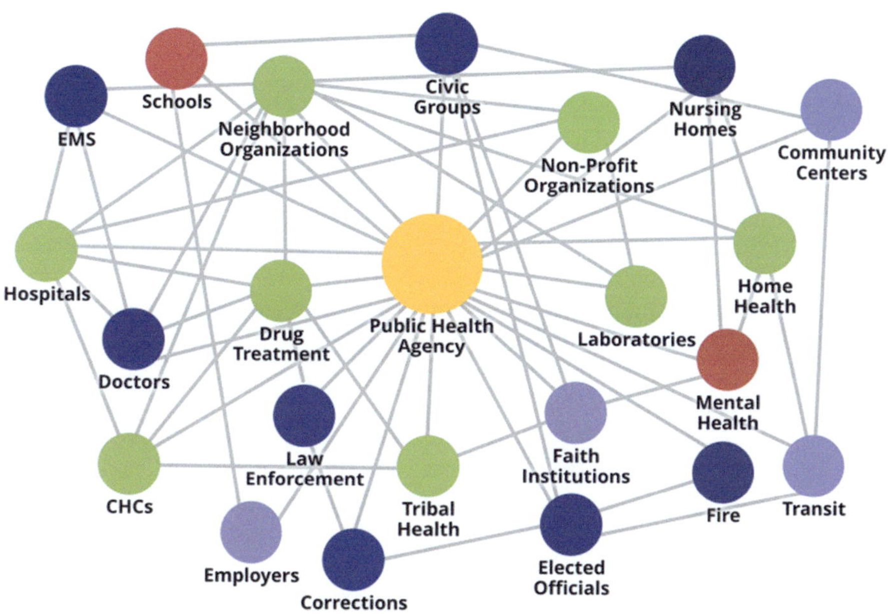

Fig. 4.2 Public health system. (with courtesy from the Centers for Disease Control and Prevention)

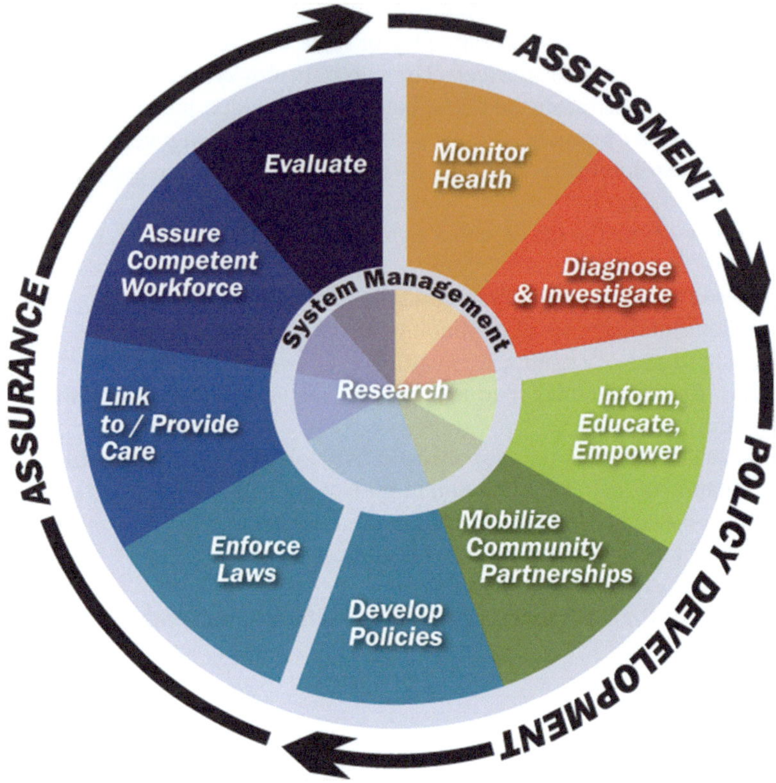

Fig. 4.3 Ten essential public health services. (with courtesy from the Centers for Disease Control and Prevention)

4.1.2 Primary Healthcare for Rural Area

Mostly, the periphery rural area of developing countries and people of least developing countries face a lot of hurdles to access primary health services. But, a variety of interpretations on rural area have been in practice to defined rural area of a specific geographical location. In general, a rural area or countryside is the location of an area that is located outside towns and cities [6].

The Health Resources and Services Administration of the US Department of Health and Human Services defines the word *rural* as all population, housing and territory not included within an urban area. Whatever is not urban is considered rural. In order to serve effectively, the Centers for Medicare and Medicaid Services (CMS), previously known as the Health Care Financing Administration (HCFA), was established. The CMS is a federal agency within the US Department of Health and Human Services (HHS). The CMC's functions as partnership with the state government to assist Medicares (it is a federal and state program that helps with medical costs for some people with limited income and resources) and the Children's

Health Insurance Program (CHIP) for effective implementation of primary healthcare services. Thus, in association with the federal state and local agencies, CHS campaigns for practicable solutions in serving 60 million Americans, and giving a best example for effective health services to rural occupants. The main targets of CMS are to improve quality care in our healthcare system, make care safe, strengthen person and family engagement and promote effective communication of care. Access to healthcare services is critical to good health, but rural residents face a variety of access barriers.

The following factors are responsible for inadequate health services in the rural area.

4.1.2.1 Barriers to Healthcare Access

Mostly, urban area of developing countries has provision of facilities for speciality healthcare services. In order to avail such facilities, the rural populations are dependent on the urban locality. But, due to the lack of proper transport services for the rural people to avail such facilities, it becomes a burden in terms of travel time and cost and the time away from the workplace. In addition, the lack of reliable transportation (mainly ambulance service) is a major barrier to care. So, the elderly residents who have chronic conditions suffer much from multiple visits to outpatient healthcare facilities.

4.1.2.2 Poor Health Literacy

To bring primary health awareness among the people, health literacy is an unavoidable practice, especially in rural locality. Basic knowledge on healthcare impacts a patient's ability to understand health information and instructions from their health councillor. In developing countries, most of the rural communities are under extreme poverty and illiteracy. With poor knowledge on health and feeling of frustration, people hesitate to express their health problem before the healthcare councillor. In order to further improve in rural healthcare service it is necessary for a good governance to mobilise at government level for providing latest technical assistance to health councillor for better healthcare service through rural communication network service.

4.1.2.3 Social Stigma and Poverty

To keep privacy is one of the main concerns in rural area, which acts as barriers for healthcare access. Generally, rural people feel reluctant to communicate health problem like seeking care for mental health, substance abuse, sexual health, pregnancy and common chronic illness due to uneasiness or privacy concerns. Mainly, the patient's confidence lies with the mutual understanding between health councillor and concern patient under care. Sometimes, the patient feels hesitant before a health councillor who used to be a family friend to discuss health issues like HIV, AIDS or any other related diseases. So, it would be a better choice to have colocation or integration of behavioural health services with primary healthcare services in the same primary health centre.

4.1.2.4 Inadequate Workforce

Due to geographical location of rural area, healthcare workforce from neighbouring urban locality feels reluctant to be in rural for service. The shortage in healthcare workforce impacts on healthcare access in rural communities. Adequate healthcare work force is needed to provide updated healthcare access at rural community level.

Shortage of public health workforce has long been a chronic problem in rural area. This is well realized by governance of a state, but due to the lack proper implementing polices and poor deputation benefits for skilled health workforce, effective rural service has been a challenging issue. In this connection, the IT sector can be helpful for a government to preach on-line health education, may be one key for solving crisis. But, several factors like timely replacement of healthcare workers in place of retard persons, increases in chronic diseases, limited facilities for education programs, and increase in aging population need special attention. Currently, at global level the shortage of skilled healthcare professionals is increasing. A 2016 study published in *Human Resources for Health* indicated a global demand for 80 million healthcare workers by 2030.

4.1.2.5 Health Insurance Coverage

Generally, the practice of health insurance is mandatory for all citizens in a developed country. Many countries including France, Dubai, the United States, Oman, Switzerland, Azerbaijan and Kazakhstan either have such schemes in placed already or are planning to implement this type of policy in 2019. Many developing countries are currently considering the possibility of introducing compulsory health insurance schemes, at least in urbanised area. But in the case of developed countries like the United States, the possibility of not having health insurance in countryside area is less. According to the US Census Bureau report (2017), without health report was about 9.2%. Individuals without health insurance have less access to healthcare services. At least half of the world's population still do not have full coverage of essential health services. About 100 million people are still being pushed into extreme poverty because they have to pay for healthcare. About 12% of the world's population pay at least 10% of their household budget for healthcare [7]. All UN member states have agreed to try to achieve universal health coverage (UHC) by 2030, as part of the Sustainable Development Goals.

The basic theme of UHC is that all individual and communities should have provision for availing health services without any financial burden, at the cost of their household expenses. The halt services scheme should be mainly for all types of quality health services, from health promotion to prevention, treatment, rehabilitation and palliative care. In addition UHC desires to ensure everyone to access the services that are linked to primary causes of disease and health and quality-based monitoring process to which the healthcare service is extended. Campaign for universal health target is one of the major goals in adopting Sustainable Development Goals in 2015. Sound healthcare increases the efficacy of children's education, gives guidance to prevent poverty and is helpful in developing sustainable rural economy.

Currently, governance of many countries is in the process of framing new plan or bringing amendments in existing health services' guidelines to follow UHC campaign for universal healthcare. To follow UHC guideline, it is essential for a nation to strengthen health systems with adequate finance provision to implement UHC, in timely manner. In rural area of a developing country, people pay a lot of money for services out of their own pockets; the poor are often unable to avail all types of health services on need base. So, pooling funds from compulsory funding sources (such as mandatory insurance contribution) can give financial relief to poor citizens, mainly from rural area. The quality and efficacy in health services is mainly dependent on the availability, accessibility and capacity of health workers to perform quality people-centred integrated care. Other decisive factors for successful implementation of universal healthcare policies are good governance, reliable and applicable procurement and well-functioning healthcare information network system.

4.1.3 Healthcare, Food Insecurity and Hunger

Hunger is not an issue for public healthcare. But without food, people's mental and physical health deteriorates. So, hunger is intimately related to public health issue. Hunger not only affects physical health, but it also affects psychological health. Mostly, health conditions are closely linked to people who aren't food secure which include asthma, diabetes, depression, obesity and arthritis. So, individuals with diet-sensitive chronic disease and food insecurity experience more difficulties managing their health, leading to more health complications. Hunger also leads to fatigue and malnourishment at all which exacerbates chronic disease, mental health and behaviour problems.

A famous phrase says "you are what you eat". So, it is necessary to consume nutritious food to keep the body and mind fit. Even in developed countries, millions suffer the soreness of hunger which ultimately affects the growth and development of a nation. It has been observed that people are 2.9 times more likely to be in poor health and have a higher likelihood of chronic conditions. Hunger has negative impact on the health of children 10–15 years of age.

4.1.4 Framework for Rural Health Technologies

Rural people have poorer standard of health as compared to urban counterparts. The rural people suffer from higher disability rates, shorter lifespan, higher infant mortality rates and higher death rates due to injuries, circulatory and respiratory diseases and diabetes as compared to urbanised area. The main causes for these mishappenings are low income, illiteracy, poverty and unemployment [7–11]. There are a number of factors associated with universal healthcare program. Social status, literacy drive, income per capita, employment opportunity, social inclusion, physical environment, women's health conditions and gender equality are some of the important factors responsible for the overall health profile of a community [10, 12], whether it is in rural or urbanised area. Mostly, from population point of view, six programmes are applicable for universal healthcare, mainly of rural areas (Fig. 4.4).

4.1.4.1 Identify a Rural Community

The Census Bureau identifies two types of urban areas, i.e. (a) urbanised areas (UAs) of 50,000 or more population and (b) urban clusters (UCs) of at least 2500 and less than 50,000 people. "Rural" defines all population, housing and territory not included within an urban area.

The Office of Management and Budget (OMB) designates areas on the basis of the Standards for Delineating Metropolitan (urbanised area >50,000 Population) and Micropolitan Statistical Area urban cluster having at least 10, 000–50,000 population.

The US Department of Agriculture Economic Research Services defines rural as open countryside or rural towns (<2,5000 people) or peripheral of urban area ranging from <2,500 to 49,999 populations.

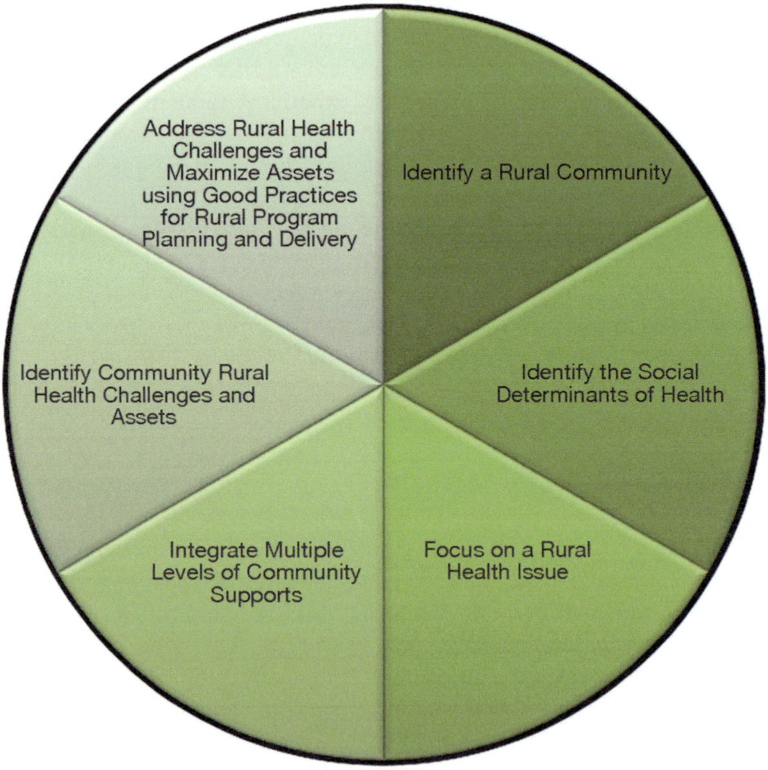

Fig. 4.4 Schematic presentation of rural health framework

4.1.4.2 Social Determinants of Health

The social and economic environment, the physical environment and the person's individual characteristics and behaviours are the key determinants of healthcare services. Income and social status, education, physical environment, social support networks, genetic, health services and gender are major factors that affect healthcare services [13, 14]. Focus on the following rural health issues:

Although provision of adequate access to healthcare services in rural areas is a critical public policy concern, still it is necessary for a good governance to design and implement viable and sustainable solution, mainly focusing on health of the rural population. Most primary health centre does not receive grant assistance in sufficient amount, in a timely manner. So, it is necessary that larger and better financed urban provider and nongovernment organisations (NGOs) should extend philanthropic attitudes, both for patients and primary care physicians which would be helpful in monitoring the decline of rural healthcare delivery system. A good governance should frame effective and efficient policies so that small hospitals must continuously expend resources for medical staff recruitment while contending with unfavourable economies of scale that make it difficult to staff for long-term continuity in service [15, 16].

Realisation of universal healthcare system has created opportunities to address rural health problems and new threats to the fragile rural health system. Bringing new amendment in the nature of healthcare delivery in rural areas will support Health Care Financing Administration (HCFA) and other developing new approaches to upgrade rural communities and providers in delivering high-quality, cost-effective services to Medicare and Medicaid beneficiaries and rural citizens.

Besides this, focus should also be given to proper channelization of financial grants received from various international volunteer organisations, health-related network services to bring awareness related to primary health management [17, 18] and the delivery of mental health services to rural Medicaid beneficiaries [19, 20] and functioning of rural managed-care entities [21–23].

4.1.4.3 Integrate Multiple Levels of Community Support

Individual effort to bring social inclusion for changing lifestyle behaviour at population level is noticed to be unsuccessful [24]. In order to bring universal healthcare at population level, it is also necessary to focus on changes at the environmental, system and policy levels and bring amendment institutional management policy system [25–27], pre-school [28], rural locality [29, 30] and development of food warehouses for instant supply [31, 32]. But relying on individual level is not promising.

So, attention has been diverted to integrated and coordinated programmes that assist multiple communities setting and bring change in policies which can be helpful at population level [33–40]. By such process multiple settings like health clinic, public entertainment park, school, primary health clinic, supermarket, etc. can be developed at population level [41, 42]. A rural population health approach uses multiple levels of support from various sectors.

4.1.4.4 Identify Community Rural Health Challenges and Assets

Mainly, rural health challenges and assets are based on observation and identification of social determinants of health [14]. Careful survey on the status and health profile of the population of rural communities provides collective information on population health. In addition, intensive research on rural population health and challenges, collective information analysis of health councillors and surveillance play important role in identifying the variable factors responsible for projecting health profile of rural population. Example of challenges is assets to healthcare services, geographic and social isolation and poverty. For a healthy social growth, assets like physical infrastructure (mainly building), green space, favourable social structure and improved agriculture practices are unavoidable factors.

4.1.4.5 Rural Health Challenges with Maximum Assets

Social determinants of health are mainly depending on how to tackle health challenges at population level with maximising assets. This approach is sort of a practical solution for developing and implementing strategy plan to improve rural population health.

4.2 Innovative Health Technologies for Rural Area

Emerging trend in innovative health technology has brought new scope for peripheral rural area of developing countries to learn about primary healthcare, sanitation and intake of clean water. Innovative healthcare technology is a modern access in rural or remote areas for keeping instant knowledge on day-to-day healthcare tips. Luckily, recent advances in telemedicine and a variety of creative trackers and mobile health tools are making this easier. The health information technologies (HIT) have potential to increase the efficacy of quality, safety and quick delivery of healthcare services in rural communities. HIT can act as an excellent bridge between super-specialised health hubs and rural primary health centre to provide updated information on latest clinical diagnosis practices. HIT is a great challenge in implementing, monitoring and managing rural health services in a timely manner. HIT uses technology to store, secure, retrieve and transfer protected health information electronically within intra-healthcare systems and community settings.

4.2.1 Components of HIT

4.2.1.1 Electronic Health and Telemedicine

Various emerging planning networks have been in the process of implementing information and communication technologies (ICT) in order to enhance the efficacy of public healthcare system. The development of ICT infrastructure is a critical factor to determine the effectiveness of rural health services. ICT can bridge distances by providing instant access to clinical knowledge, specialised expertise and services not available in isolated areas.

Two important trends have been emerging to upgrade healthcare delivery system. The increased magnitude of healthcare system has been prominent due to private investments in clinical research which resulted in the development of new drugs, medical devices and other interventions, opening new scope to improve health and get quick relief from suffering. But translating new medical knowledge into practice is not encouraging [43]. ICT-aided information can minimise time and efforts required to translate knowledge effectively into practice and safely utilise the newly developed drug and medical devices for rural health services [44, 45]. Secondly, the life expectancy in developed countries has been increased, mostly due to the successful implementation of healthcare system. For example, in America the life expectancy has been in increasing trend, since the last two decades [46–49].

By the application of information technology, we can bring amendment in life profile of rural inhabitants. Generally, rural health councillors work in more isolated rural area with limited population density and are trying to assess the effect of information technology. So, it is necessary to coordinate a larger number of transfers to facilities in distant location [50–52]. So, improving the quality of healthcare ultimately requires improving the availability of healthcare information [53].

(i) *Electronic health records:* An electronic health record (EHR) is a digital device for recording health information of a patient under treatment. EHR provides service to an authorised health councillor or doctor for the detailed information of patient's health profile which include the following: (a) allow access to patient medical history, diagnoses, medications, treatment plans, immunisation dates, allergies, radiology images and laboratory test results; (b) allow access to evidence-based tools that providers can use to make decisions about a patient's care; and (c) automate and streamline provider workflow. One of the key features of an EHR is that health information can be created and managed by authorised providers in a digital format capable of being shared with other providers across more than one healthcare organisation.

(ii) *Computerised physician order entry (CPOE):* CPOE is also referred to as computerised provider entry or computerised provider order management (CPOM). It is a process of electronic entry of medical practitioner instructions for the treatment of patients (particularly hospitalised patients) under his or her care. The benefits of CPOE include safer and more consistent patient-centred care that is lasting and measurable. This system supports better patient safety and quality care.

4.2.2 Future Prospects of Innovative Healthcare Technology

The future of healthcare is progressively changing with the advancement of medical and health sciences. Data science and machine learning has catalysed healthcare services, especially in developing countries. The key opportunities with the development of innovative healthcare technology include health monitoring and diagnosis; medical treatment and patient care; pharmaceutical research and development; and clinical performance optimization.

According to the Institute of Medicine report (2000), handwritten medical reports, manual order entry, and non-standard abbreviations' poor legibility resulted in erroneous expression of healthcare reports which may ultimately be responsible for the death of patients.

About two decades back, the US Institute of Medicine reported on the basis of multiple studies by variety of organisations that about 44,000–98,000 people die in United States due to preventable medical errors [54, 55].

So, the new health system for the twenty-first century advised rapid adaptation of electronic patient records and electronic medication ordering by developing modernised infrastructure for IT to support clinical decision.

Implementation of electronic medical record (EMR) systems will lead to major healthcare savings, reduce medical error and improve health. But, there has been poor progress towards achieving these benefits. For example, even in developed country like the United States, the use of EMR system is about 15–29% [56–58]. The main hurdle is high cost and expensive maintenance.

4.3 Manual for Healthcare

In order to develop global harmonised healthcare practice at universal level, the World Health Organization (WHO) framed guidelines for infection prevention and control (IPC) programmes to have quality and safety of health service for world population. The entire project is aimed to improve efficacy in quality healthcare services' delivery system and the health outcomes of the people who avail those services. The entire health services' manual consists of eight key areas of IPC. The principles and guidance mentioned in this manual are applicable to global population with minimum favourable resources. The eight contents of eight areas of IPC include (1) IPC programmes; (2) evidence-based guidelines; (3) education and training; (4) healthcare-associated infection (HAI) surveillance; (5) multimodal strategies; (6) monitoring and audit of IPC practices and feedback; (7) workload, staffing and bed occupancy; and (8) built environment, materials and equipment for IPC. This manual provides practical guidance, and with the quotation of world data to make the user convenience and clear in clinical use. It also gives emphasis on the development of a sustainable action plan which can be made easier for the local health councillor while in practice. This manual can be immensely helpful in integrating and embedding IPC within the day-to-day structure and activities of healthcare facility. The significant features of this manual are:

- Sequential implementation of IPC programmes for the upgradation of population healthcare services
- Implementation of new amendments in the IPC component for bringing improvement in personal tax benefit, in a health care facility

- Briefing of common barriers and practical solution while implementing the core programmes expressed in the healthcare manual
- Convenient implementation of IPC programmes at global population level
- Supporting various aspects of IPC action plan like planning, advocacy (developing a convincing narrative to win heart and mind), assessment, guideline development, stakeholder engagement and training and education

The manual is not aimed at forceful implementation but of providing a stepwise approach to implement based on the evidence and experience of what has worked in a number of settings and introduces examples and ideas from healthcare facilities around the world which can be used by IPC teams within healthcare facilities.

In addition this manual provides practical guidance while implementing IPC programmes in the most effective way and with higher quality by minimising the risk of microbial transmission and healthcare-associated infections, including the spread of microbial resistance. This manual will be immensely helpful to nurses and health councillors where IPC programmes are nonfunctional. It can be useful to primary healthcare centres while reviewing the exits healthcare programme linked to quality improvement, patient safety, health facility accreditation and public health control.

4.4 Healthcare Access in Rural Community

Good governance desires to have homogeneity in healthcare services both in urbanised areas and least developing countries, irrespective of their economic development and infrastructure availability. Approximately, one half of the global population lives in rural areas, but these areas are served by only 38% of the total nursing workforce and by less than a quarter of the total physician workforce. Due to geographic location, the peripheral parts of developing countries and people of least developing countries have been facing problem on the proper access to healthcare. Rural health services is of multidimensional in nature covering geography, midwifery, nursing, sociology, social inclusion, economics and telemedicine like programme. The most significant aspects of access to health are overall physical, social and mental health status; disease prevention; clinical diagnosis; normal lifestyle of living; and life expectancy. But, rural can't be defined from international point of view with some standard concept [59, 60]. However, based on population density [61, 62], distance from an urban centre and geographic location, rural areas are identified [63]. One of the major problems in rural area is life expectancy. In general, life expectancy rates in urbanised area are comparatively higher than rural area [64–66]. This could have been due to poor economic conditions and inadequate facilities in health education in rural areas [67]. However, there are countries such as the United Kingdom where life expectancy in rural areas is higher than the urban areas. This may be due to the provision of better education, especially in rural health, smaller economic disparities in rural area and migration of wealthy and educated individuals to rural areas [68].

4.4.1 Health Determinants

Adjunct locality to developing countries and least developing countries receive fewer provisions for healthcare services. These areas have less preventative healthcare (mainly of mental health programmes, primary healthcare) and experience delay in attending emergencies. The inadequate availability of healthcare councillors has resulted in unconventional ways of delivering healthcare to rural dwellers. So, the good governance in some of the rural states tries to depute doctors and nurses with additional package for rural services in order to increase the efficacy in rural healthcare services [69]. For example, a rural occupant in Canada avails the same healthcare facilities as received by urbanised people. This rural healthcare facility is even extensible to the rural locality situated around a distance of 10 km periphery to urbanised locality [70], even though with inadequate healthcare facility like dentists, dental surgeons, ambulance service, blood and urine testing services [71].

In developing country like China, the central government has overall responsibility for national health legislation, policy and administration. Every citizen of China is entitled to avail the benefit of basic healthcare services being supported by the local government. In 2014, China healthcare budget was 5.5% of its gross domestic product. But, in India the state government has responsibility to develop budget for healthcare, rather than the central government [72, 73].

Rural area in India has inadequate medical professionals [74]. This is a major issue for rural access to healthcare. Due to the lack of sufficient budget, insufficient housing, lack of education for children, poor availability of drinking water, lack of continuous electricity, shortage of infrastructure and shortage of public transport with the state government, the drive for rural health in India is not an encouraging sign [75, 76]. The most unfortunate factor is the high mortality risks before the age of 5 are greater for children living in peripheral rural areas [77]. Complete immunisation practice in rural area is about 39%, and in urbanised area it is about 58% [77]. These inequalities could have been due to factors such as socio-economic status and caste. The caste issue serves as a social determinant of healthcare in India [77].

Conditions of access to healthcare in least developing countries are more horrible as compared to the rural areas of developing countries. In Africa, due to financial crunch, the government faces greater difficulties in recruiting and retaining qualified and skilled professionals in the healthcare field [77]. Even urbanised Zambia has 20 times more doctors than the rural area. Overall healthcare conditions in many rural countries like Malawi, Burkina Faso, South Africa, is unsatisfactory due to the poor ratio of doctors to inhabitants [78].

Due to geographical location, distance from urbanised areas and insufficient electricity availability, most of the rural youth are unemployed. Mostly, manual labour opportunity is available, depending on crop harvesting or crop postharvesting periods. Farming, forestry, fishing and home industry are also few opportunities available for rural habitants [79]. These occupations are often associated with health hazards and are unsecure due to unhygienic working conditions, exposure to chemicals and gases, long working hours and harsher climates [80, 81].

Living style and microenvironment surrounding the community supposed to have impact on health and longevity of individuals in rural areas. Mostly, a maximum number of rural people have the habit of smoking and poor intake of nutritious fruits and vegetables [64].

4.4.2 Global Awareness on Rural Healthcare

Earlier, around the mid-1980s, people started realising about the big gap in healthcare outcomes between individuals in rural areas and those in urban areas. Since the last two decades, there has been increased funding by government organisation and nongovernment organisation to upgrade rural healthcare services in order to increase the efficacy of the working system in implementing the concept of universal healthcare [82–84]. Some good governance wants to incorporate rural healthcare in national policy [85]. Many research organisations and NGOs such as the Centre for Rural and Northern Health Research, Laurentian University; the Center for Rural Health, University of North Dakota; and the National Rural Health Alliance have been emerging for universal healthcare campaigns [86]. Canada is one of the leading nations to decentralise primary healthcare and move towards a more regional approach. A long time back, in 2007 Canada has established Local Health Integration Network in Ontario to provide quality healthcare services. In developing country like China, the government is in the process of providing basic healthcare for its entire citizen by 2020. The major components of China's healthcare reform plan include expansion of basic medical insurance programmes and creation of a cooperative medical system, establishing a national essential drug system that encompasses drug selection, production and supply, clinical applications, and medical insurance reimbursement, and primary medical care infrastructure.

Constitutionally state governments of India have to bear the entire responsibility of healthcare rather than the central federal government. Each state government has to develop fund to meet the expense related to health services [87, 88]. The National Health Policy was endorsed by the Parliament of India in 1983 and updated in 2002 and then in 2017. In practice, the private healthcare sector takes the responsibility of healthcare service. For the same purpose, the private healthcare sector charges the money to the patients rather than to the government. The patients have to get the money either from an insurance company or from their own pocket [89, 90].

4.5 Rural Health Projects

The main objective of rural health project is to increase equitable access to quality health services with well-planned budgetary provision. In addition attention is given to communication, logistic services and generation of adequate staff such as doctors, nurses and general staff [91]. Still, many rural health projects being operated in poor areas are lacking access to basic medical help like clinics, or doctors use nontraditional methods in providing healthcare [92, 93]. In order to support and guide common people, especially of rural areas, nonprofit health information and health education source like *Hesperian Health Guides* book is made available for them to acquire information on primary healthcare and other health-related problems. As a result, even without the help of doctors or health councillors, common people can understand about primary healthcare [94].

The *Hesperian Health Guides* book was first published in 1973 and updated every few years (1, 2). The book is available in more than 80 languages and has taken credit as the most widely used public health manual in the world [95].

"Sure Start" is an ambitious project, launched in 2005 in India to improve health behaviours and increase the use of health services. The project has reached nearly 25 million people, about 2% of India's billion-plus population with services aimed at making children safer and babies healthier. Sure Start project has showed positive impact on health behaviours and increases in the use of health services like prenatal care, the number giving birth in hospitals and the number adopting recommended breastfeeding practices [96].

The main target of The Rural Health Advocacy Project (RHAP) is to bring amendments or develop healthcare project to access high quality, comprehensive healthcare services in rural areas with the aim of improving the health of the South Africa population. The RHAP is a partnership between the Wits Centre for Rural Health, the Rural Doctors Association of Southern Africa (Ru DASA) and SECTION 27 (previously the Aids Law Project), based in Johannesburg, South Africa.

The Rural Health Workforce Australia (RHWA) is a nongovernment organisation having target to solve the problem of the shortage of healthcare workers. It is funded by the Australian Government Department of Health and Ageing. The main target of RHWA is to provide accessible, quality healthcare services in rural and remote Australia.

The US Health Resources and Services Administration funds the Rural Hospital Performance Improvement Project to improve the quality of care for hospitals fewer than 200 beds. The Office of Rural Health Policy (ORHP) is attached to the Health Resources and Services Administration (HRSA) of the US Department of Health and Human Services (HHS). It is the primary federal agency for improving access to healthcare services for remote rural localities. They also train health professionals and improve systems of care in rural community. Through these agencies 60 million inhabitants of rural America get sufficient and accessible healthcare. In addition, effective management of existing healthcare institutions that aid in the recruitment and retention of physicians in rural hospital and clinics is being planned and operated by these agencies. The HRSA and ORHP manage to utilise telehealth programme to link isolated rural practitioners to medical institutions over great distances.

Hunger is one of the most important complementary factors for maintenance of normal health. So, it is foremost important to accomplish "Health for All" project with agricultural development. The United Nations food agency warned that if investment in agricultural research is not increased, especially in Asia, global efforts to achieve zero hunger will fall short and ultimately affect universal healthcare drive. Nearly 800 million hungry and undernourished people in the world are going to develop intense healthcare problem globally. The FAO estimates that the world needs to produce about 60% more food to feed a hungry world by 2050 which may ultimately reduce healthcare problem significantly.

References

1. Declaration of Alma-Ata (1978) http://www.who.int/hpr/NPH/docs/declaratio_almaata
2. Secretariat, WHO (2011) International conference on primary health care, Alma-Ata: twenty-fifth anniversaries. Report by the Secretariat. WHO
3. Cueto M (2004) The origins of primary health care and selective primary health care. Am J Public Health 94(11):1864–1874
4. White F (2015) Primary health care and public health: foundations of universal health systems. Med Princ Pract 24(2):103–116
5. Bulletin of the World Health Organization (2008) China's village doctors take great strides. World Health Organization 86(12):914–915
6. Word Net Search (2013) Wordnetweb.princeton.edu
7. Desmeules M et al (2006) How healthy are rural Canadians? An assessment of their health status and health determinants. Canadian Institute for Health Information, Ottawa
8. Hart G, Larson E, Lishner D (2005) Rural definitions for health policy and research. Am J Public Health 95:1149–1155
9. Public Health Agency of Canada (PHAC) (2008) Health topics: rural health. Retrieved August 5, 2008. http://www.phac-aspc.gc.ca/ccdpc- cpcmc/topics/rural_e.html
10. Romanow R (2002) Building on values: the future of health care in Canada. Commission on the Future of Health Care in Canada, Saskatoon
11. Smith KB, Humphreys JS, Wilson MG (2008) Addressing the health advantages of rural populations: how does epidemiological evidence inform rural health policies and research? Aust J Rural Health 16:56–66
12. Public Health Agency of Canada (PHAC) (2006) Information: determinants of health. Retrieved October15, 2012. http://www.phac-aspc.gc.ca/ph sp/phdd/determinants/determinants.html#income
13. Public Health Agency of Canada (PHAC) (2002) Population health: an integrated model of population health and health promotion. Public Health Agency of Canada (PHAC), Ottawa
14. Public Health Agency of Canada (PHAC) (2011) What determines health? http://www.phac-aspc.gc.ca/ph-sp/determinants/index-eng.php
15. Office of the Inspector General (1993) Trends in rural hospital closure: 1987–1991. Department of Health and Human Services, Atlanta. Pub. No. OEI-04-92-00441
16. U.S. General Accounting Office (1991) Rural hospitals: federal efforts should target areas where closures would threaten access to care. Washington, DC. Pub. No. GAO/HRD-91-41
17. Wellever AL, Rosenberg S (1993) Alternative models for organizing and delivering health care services in rural areas, volume 2 – state-sponsored alternative models: the state government perspective. National Rural Health Association, Kansas City
18. Office of Technology Assessment (1990) Health care in rural America. U.S. Government Printing Office; U.S. Congress, Washington, DC
19. Wagenfeld M, Murray J, Mohatt D, DeBruyn J (1994) Mental health and rural America: 1980–1993. Office of Rural Health Policy, Rockville
20. Christianson J, Moscovice I (1993) Health care reform: issues for rural areas. University of Minnesota Rural Health Research Center, Minneapolis
21. Nycz GR, Wenzel FJ, Freisinger RJ, Lewis RF (1987) Medicare risk contracting: lessons from an unsuccessful demonstration. JAMA 257(5):656–659
22. Rosenberg and Associates (1995) Issues and options facing rural communities under managed care. Point Richmond, Richmond

23. Weisgrau S, Rosenberg S (1993) Alternative models for organizing and delivering health care services in rural areas, volume 3 – state-sponsored alternative models: experiences of participating providers. National Rural Health Association, Kansas City
24. Maziak W, Ward KD (2009) From health as a rational choice to health as an affordable choice. Am J Public Health 99:2134–2139. https://doi.org/10.2105/AJPH.2008.155382
25. Campbell K, Waters E, Meara SO, Summerbell C (2001) Interventions for preventing obesity in childhood. A systematic review. Obes Rev 2(3):149–157
26. Foster GD et al (2008) A policy-based school intervention to prevent overweight and obesity. Pediatrics 121:794–802. https://doi.org/10.1542/peds.2007-1365
27. Moore GF et al (2014) Impacts of the primary school free breakfast initiative on socio-economic inequalities in breakfast consumption among 9–11-year-old schoolchildren in Wales. Public Health Nutr 17:1280–1289. https://doi.org/10.1017/S1368980013003133
28. O'Connell ML et al (2012) Repeated exposure in a natural setting: a preschool intervention to increase vegetable consumption. J Acad Nutr Diet 112:230–234. https://doi.org/10.1016/j.jada.2011.10.003
29. Eklund L et al (2011) Workplace health promotion in the Nordic countries – a review of literature 1986–2008. Glob Health Promot 18:15–22
30. Thorsen AV et al (2010) Long-term sustainability of a worksite canteen intervention of serving more fruit and vegetables. Public Health Nutr 13:1647–1652
31. Gittelsohn J (2010) An urban food store intervention positively affects food-related psychosocial variables and food behaviours. Health Educ Behav 37:390–402
32. Holmes AS et al (2012) Effect of a grocery store intervention on sales of nutritious foods to youth and their families. J Acad Nutr Diet 112:897–901
33. Satterfield DW et al (2003) Community-based lifestyle interventions to prevent type 2 diabetes. Diabetes Care 26:2643–2652
34. Verheijden MW, Kok FJ (2005) Public health impact of community-based nutrition and lifestyle interventions. Eur J Clin Nutr 59:S66–S75
35. Economos CD et al (2007) A community intervention reduces BMI z-score in children: shape up Somerville first year results. Obesity 15:1325–1336
36. Ganann R et al (2012) Community-based interventions for enhancing access to or consumption of fruit and vegetables among five to 18-year olds: a scoping review. BMC Public Health 12:711
37. Lv J et al (2014) Community-based multilevel intervention for smoking, physical activity and diet: short-term findings from the community interventions for health programme in Hangzhou. China J Epidemiol Commun Health 68:333–339
38. Van Koperen TM et al (2013) Characterizing the EPODE logic model: unravelling the past and informing the future. Obes Rev 14:162–170
39. Economos CD (2013) Shape up Somerville two-year results: a community-based environmental change intervention sustains weight reduction in children. Prev Med 57:322–327
40. Novotny R (2013) The Pacific way to child wellness: the children's healthy living program for remote underserved minority populations of the Pacific region (CHL). Hawaii J Med Public Health 72:406–408
41. Collie-Akers VL (2007) Analyzing a community-based coalition's efforts to reduce health disparities and the risk for chronic disease in Kansas City. Missouri Prev Chronic Dis 4:A66
42. Nastasi BN, Hitchcock J (2009) Challenges of evaluating multilevel interventions. Am. J. Community Psychol 43:360–376
43. Balas EA et al (1998) Effect of linking practice data to published evidence: a randomized controlled trial of clinical direct reports. Med Care 36(1):79–87
44. Becher EC, Chassin MR (2001) Improving quality, minimizing error: making it happen. Health Aff 20(5):164–179
45. Jerome RN et al (2001) Information needs of clinical teams: analysis of questions received by the clinical information consult service. Bull Med Libr Assoc 89(2):177–185
46. Anderson JG, Horvath J (2002) Chronic conditions: making the case for ongoing care. Johns Hopkins University and The Robert Wood Johnson Foundation, Baltimore

47. IOM (2001) Crossing the quality chasm: a new health system for the 21st century. National Academy Press, Washington, DC
48. IOM (2003) Priority areas for national action: transforming health care quality. The National Academies Press, Washington, DC
49. NCHC (National Coalition on Health Care) (2002) Accelerating change today. National Coalition on Health Care, Washington, DC
50. Melzner SM, Grossman DC, Hart LG, Rosenblatt RA (1997) Hospital services for rural children in Washington state. Pediatrics 99(2):196–203
51. Hart LG, Salsberg E, Phillips DM, Lishner DM (2002) Rural health care providers in the United States. J Rural Health 18:211–232
52. Rosenfeld BA et al (2000) Intensive care unit telemedicine: alternate paradigm for providing continuous interventist care. Crit Care Med 28(12):3925–3931
53. James B (2003) Information system concepts for quality measurement. Med Care 41:I-71–I-79
54. Ali M, Marks J, Stroup D, Gerberding J (2000) Actual causes of death in the United States, 2000 (PDF). JAMA JAMA 291(10):1238–1245
55. Ballweg R (2013) Prevention of medical errors. Physician assistant: a guide to clinical practice, 5th edn. Elsevier/Saunders, Philadelphia. ISBN:978-1455706570
56. Taylor H, Leitman R (2005) European physicians, especially in Sweden, Netherlands, and Denmark, lead U.S. in use of electronic medical records, Harris Interactive 2(16) (8 August 2002), www.harrisinteractive.com/news/newsletters_healthcare.asp
57. Fonkych K, Taylor R (2005) The state and pattern of health information technology adoption. RAND, Santa Monica
58. Taylor R et al (2005) Promoting health information technology: is there a case for more-aggressive government action? Health Aff 24(5):1234–1245
59. Demographic and Social Statistics (2017) Population density and urbanization. United Nations Statistics Division. https://www.Unstatus.un.org
60. Pong RW, Pitbaldo RJ (2001) Don't take "geography" for granted! Some methodological issues in measuring geographic distribution of physicians. Can J Rural Med 6:105
61. du Plessis V et al (2001) Definitions of rural (PDF). Rural and small town Canada analysis bulletin. 3 (3)
62. United-States Department of Agriculture (2007) Rural definitions: data documentation and methods
63. Pitblado JR (2005) So, what do we mean by "rural", "remote" and "northern"? Can J Nurs Res 37(1):163–168
64. Ottawa: Canadian Institute for Health Information (2006) How healthy are rural Canadians? An assessment of their health status and health determinants
65. Australian Institute of Health and Welfare (2005) Rural, regional, and remote health: indicators of health, ISBN 9781740244671
66. Shen J (1993) Analysis of urban-rural population dynamics of China: a multiregional life table approach. Environ Plan 25(2):245–253
67. Stephens S (2014) Gap in life expectancy between rural and urban residents is growing. Center for Advancing Health
68. Ramesh R (2010) Country dwellers live longer, report on 'rural idyll'
69. Rourke J (2008) Increasing the number of rural physicians. Can Med Assoc J 178:322–325
70. Ng E et al (1999) How far to the nearest physician. Rural Small Town Anal Bull 1:1–7
71. Halseth G, Ryser L (2006) Trends in service delivery: examples from rural and small town Canada, 1998 to 2005. J Rural Community Dev 1:69–90
72. National Health Policy (2017) Ministry of Health and Family Welfare. Jugal Kishore (2005) National health programs of India: national policies & legislations related to health. Century Publications
73. Thayyil J, Jeeja MC (2013) Issues of creating a new cadre of doctors for rural India. Int J Med Public Health 3(1):8
74. Balarajan Y, Selvaraj S, Subramanian SV (2011) Health care and equity in India. Lancet 377(9764):505–515

75. Costa D, Al-Muniri A, Diwan A, Vinod K et al (2009) Where are healthcare providers? Exploring relationships between context and human resources for health Madhya Pradesh province, India. Health Policy 93(1):41–47
76. Rama B et al (2010) Inequities in access to health services in India: caste, class and region. Econ Polit Wkly 45(38):49–58
77. The World Bank (2008) Health workers needed: poor left without care in Africa's rural areas
78. Bring Health Care Services to Rural Africa (2018) The Atlantic philanthropies, rural health, Wikipedia
79. Bollman RD (2008) An overview of rural and small town Canada. Can J Agric Econ 39(4):805–817
80. Gerberich SG et al (1998) Machinery-related injuries: regional rural injury study-I (RRIS-I). Accid Anal Prev 30(6):93–804
81. Pickett W et al (1999) Fatal work-related farm injuries in Canada, 1991–1995. Canad Medi Associ Jr 160(13):1843–1848
82. Obama B (2009) A new era of responsibility (United States Office of Management and Budget), https://en.wikisource.org
83. Humphreys J et al (2002) Whither rural health? Reviewing a decade of progress in rural health. Aust J Rural Health 10(1):2–14
84. Rural proofing guidance (2013) Department for Environment, Food & Rural Affairs. Government of the United Kingdom, www.nationalarchives.gov.uk/doc/open-government-licence/version/3/ or email
85. Rural Proofing (2019) New Zealand different. Ministry for Primary Industries. Government of New Zealand. https://www.mpi.govt.nz
86. World Health Organization (1986). Ottawa Charter for Health Promotion. First International Conference on Health Promotion
87. National Health Policy (2017) Ministry of health and family welfare
88. Kishore J (2005) National health programs of India: national policies & legislations related to health. Century Publications. ISBN 978-81-88132-13-3
89. Berman P (2010) The impoverishing effect of healthcare payments in India: new methodology and findings. Econ Polit Wkly 45(16):65–71
90. Britnell M (2015) In search of the perfect health system. Palgrave, London, p 60
91. Strasser R (2003) Rural health around the world: challenges and solutions. Fam Pract 20(4):457–463
92. Weisgrau S (1995) Issues in rural health: access, hospitals, and reform. Health Care Financ Rev 17(1):1–14
93. Babu E (2010) Where there is no doctor. JAMA 303(9):885
94. Hebert E et al (2016) mBody health: digitizing disabilities in Sierra Leone. 2016 IEEE Global Humanitarian Technology Conference (GHTC): 717–724
95. Hesperian Health Guides (2008) Health of women with disabilities: an interview with Jane Maxwell of the Hesperian Foundation, Moving Toward Inclusion, Mobility International USA, WikiVisually, https://wikivisually.com
96. Arnab A et al (2014) Evaluating a large-scale community-based intervention to improve pregnancy and newborn health among the rural poor in India. Am J Public Health 105(1):144–152

Rural Women Encounter Hunger and Poverty

5

About 1.5 billion women live in extreme poverty. Still, rural women play a significant role in managing households and communities by extending their support and cooperation in achieving food and nutrition security, generating income and upgrading the standard of rural livelihood and overall well-being. They also significantly catalyse the agricultural practices and development of small-scale agro-industry and monitor local and global economies. The parochial social norms and practices prevent them from many social inclusions and make them completely handicap in exercising their right for social justice. Still, they are also an important target group for the MDGs and SDGs' drive to bring a sustainable pattern of life free from social agony, poverty, malnutrition and hunger. But in spite of their so much crucial efforts in assuring food security, rural women confront many social barriers.

Social exclusion for women has been a chronic practice in rural areas of developing and least developing countries. Most of the girls do not get opportunity in formal education, are restricted in working outside the household and get low daily wages for manual labour both in the agriculture fields and rural home industry. However, the implementation of MDGs and SDGs, in some part of the rural area, changes the lifestyle of women and girls at various levels of social activities like household management, local agriculture practices, literacy drive and volunteering in primary healthcare services.

5.1 Women and Hunger

Hunger and poverty are complementary to rural women, mostly from peripheral regions of developing countries and least developing countries. But rural women have potential to defeat hunger. Women have diversified role in the society as mothers, farmers, teachers and entrepreneurs. In order to have compactable and healthy family, good nutrition is of primary importance. Poor nutrition has negative impact on women, which ultimately affects the health of their children. The children of

B. K. Behera et al., *Move Towards Zero Hunger*,
https://doi.org/10.1007/978-981-32-9800-2_5

malnourished women are more prone to higher rates of disease and infection with higher mortality rates [1].

In pregnancy, menstruation and lactating period, women are more susceptible to malnutrition. During this period women should have adequate amount of energetic food rich in proteins and nutraceuticals [2]. About 60% of women and girls are undernourished. At least 120 million women in developing countries are underweight [1]. As per UNICEF 2014 report [3], babies suffering from low birthweight (LBW) are from malnourished mothers. In 2015, 20.5 million newborns, an estimated 14.6% of all babies born globally that year, suffered from LBW. These babies were more likely to die during their 1st month of life [1, 4]. So, it is necessary to have healthy food and nutritional fruits and vegetables. The World Health Assembly (WHA) targets to reduce LBW by 30% between 2012 and 2025.

Obesity also resulted due to malnutrition, which may bring risks such as labour complications and poor child health [5]. The existence of gender inequality in rural areas is also responsible for hunger-stricken areas of developing countries [6].

5.1.1 Malnutrition

Malnutrition is defined as deficiencies, excesses or imbalances in food intake.

Hunger and malnutrition are the most serious problems in rural women, especially for poverty-stricken areas. Women in the reproductive age group and children are most vulnerable to malnutrition due to low dietary intakes, uneven availability of food within the household, non-availability of food storage systems and lack of sanitary management in food preparation areas. In addition highly expensive nutritional diet is disgusting for pregnant and lactating rural women. Various forms of malnutrition include undernutrition (wasting, stunting, underweight), inadequate vitamins or minerals, overweight, obesity and resulting diet-related non-communicable diseases.

As reported by the WHO in 2018 [7], 1.9 billion children are overweight, while 462 million are underweight. The data on wasted children (52 million) and severely wasted (17million), stunted growth (155 million) and overweight (41 million) are alarming, even after the lapse brought about the implementation of SDG program. The low- and middle-income countries show 45% death per year due to malnutrition. The developmental, economic, social and medical impacts of the global burden of malnutrition are serious and lasting, for individuals and their families, for communities and for countries.

The term malnutrition addresses three broad groups of conditions.

5.1.1.1 Protein-Energy and Vitamin Deficiency Malnutrition

Various dietary problems are associated with malnutrition, especially with rural women. Deficiency in protein diet causes impact on strengthening the health of rural women, especially before and after childbirth. Proteins are complex molecules composed of different amino acids. Our body's essential amino acids are available from the diets since they are not synthesised in the human body. Proteins are functional components of every living cell. Body muscles, cartilage and skin are rich in proteins. Mainly, malnutrition in adult is due to insufficient food, infections, malabsorption of nutrients and eating disorders. Mostly, rural pregnant women due to lack of insufficient portentous diet suffer from poor development of foetus. Mostly, complication in pregnancy is due to poor intake of essential proteins which leads to problems like maternal morbidity and mortality, neonatal mortality and low infant birthweight. A malnourished woman faces the problem of secretion of adequate amount of breast milk, which can cause malnourishment in infants.

Micronutrient malnutrition is defined as deficiency of vitamins or mineral leading to health disorder (disease). About two billion people of the total world population are suffering from micronutrient malnutrition. The main cause of preventable blindness in children is vitamin A deficiency (VAD). Deficiency of vitamin A and also night blindness in pregnant women may increase the risk of maternal mortality. People (children and pregnant women) from Southeast Asia and Africa are mainly victims of VAD. An estimated 250,000 to 500,000 vitamin A-deficient children become blind every year [7, 8].

5.1.1.2 Iron Deficiency

A pregnant woman needs enough iron as the quantity of blood increases to support the developing baby. The deficiency of iron causes anaemia which may increase the risk of life. Mostly, women in developing countries and least developing countries are under the grip of iron deficiency and suffering from anaemia [9–12]. Children born to iron-deficient mother are more likely to have impaired physical and mental development and inadequate immune potential.

5.1.1.3 Undernourishment

Malnutrition is also caused by undernourishment where an individual is not getting adequate quantity of enough calories, protein or micronutrients [13, 14]. Mainly the physical and mental disorders of children are caused by undernourishment of pregnant women [15]. Extreme undernourishment (starvation) of pregnant women also causes short height, thin body, poor energy level and swollen legs and abdomen of children [13, 15]. The problem of undernourishment is mostly prevailing in rural areas due to inadequate availability of quality food, from time to time. This could have been due to high food price and poverty. Lack of breast feeding leads to a number of ill health problems like infection of diseases (gastroenteritis, pneumonia, and malaria), risk of childhood obesity, type 1 and type 2 diabetes, leukaemia and sudden infant death syndrome. In some developing countries, both overnutrition and undernutrition lead to the problem of obesity [14]. Other causes of malnutrition include *anorexia nervosa* and *bariatric surgery* [15, 16].

In 1990 the number of undernourished people in the world was about 997 million, and by 2017 it has been reduced by 176 million [17, 18]. In 2015, protein-energy malnutrition has resulted in about 323,000 death cases as compared to the death toll of 510,000 in 2012 [19, 20].

5.1.1.4 Resource Use and Malnutrition

Due to social exclusion, rural women are not in a position to use available resources for family welfare, to improve the living standard, and many other household works. When women have more opportunity to access resources, their potential to decide how to use them could be immensely helpful in maternal nutrition improvement, child health improvement and microeconomic growth at community level [21].

In the peripheral parts of developing countries and least developing countries, rural women have to perform many types of unpaid domestic work under family obligation and social constraints. For example, in order to meet the day-to-day energy demand, they have to go to distant forest areas to fetch firewood, at the cost of burning heavy kilocalorie energy. But not in financially so much sound enough to supplement the energy burned by quality diet. In addition, rural women have lack access to resources needed to manage better food security which would be helpful in adverse climatic conditions. These resources include land and financial assistance and livelihood training. Mostly women prepare food for the family members, but the social custom forces them to eat lastly. This means they have less access to sufficient quantity of food. So, this sort of trend of inequality in food consumption within family is a key issue leading to malnutrition in women, especially in rural areas.

5.1.1.5 Malnutrition Due to Natural Calamity and War

Often, natural calamity and warlike crisis lead to social separation of family and concern members. As a result, especially women and children face horrible condition while leading livelihood. While fleeing war or while in refugee camp, pregnant and lactating women will be extremely vulnerable under such unfavourable crisis. Due to extreme lack of proper quality food in adequate amount and poor medical care, women and children severely suffer from malnutrition problem. Women and children in fragile setting (the regions affected by violent or natural disaster) are more prone to causality and disease.

5.2 Women and Poverty

Many of the world's most poor are women. Thus, poverty eradication is the foremost priority for good governance. The World Bank defined "extreme poverty" as living on less than $1.90 a day (Fig. 5.1a, b). For middle- and high-income countries, it is $3.20–$5.50 and $21.70 a day, respectively. The United Nations hope to eradicate extreme poverty by 2030 as part of its Sustainable Development Goals [22].

Generally, poor availability of access for rural women prevents them to avail opportunity to spend livelihood under zero hunger and reduced poverty. Rural women remain busy at maximum in spending time collecting clean water from distant water resources, gathering firewood from nearby forest, caring for children and sick and preparing food for family members, as compared to urban women. In rural areas of developing countries like India, women spend more than twice as much time fetching firewood and water to drink (Fig. 5.1a, b).

Fig. 5.1 (**a**) Gathering of rural women to fetch water from a well, nearby a rural village. (**b**) Rural village women collecting firewood from a nearby forest in India

According to recent reports from the UNICEF and WHO, about 263 million people worldwide have access to safe water sources but need to spend at least 30 min walking or queuing to collect their water for drinking [23].

Job opportunity for rural women is extremely overlooked. Many of the jobs available in rural areas are on a short-term basis, depending on seasonal requirement, more precarious and less protected than those of rural men and urban people. The rural women working group in agriculture sector is usually equivalent to or higher than the male counterpart. About 70% in South Asia region and 60% in sub-Saharan working class are in agriculture, as compared to male workers [24].

Rural women participation in agriculture as unpaid or as freely contributing family workers or in marketing rural goods in nearby village market is on a no-loss no-profit basis. This is mainly due to inefficiency of good governance in framing rural trade policies and developing provision for logistic system for rural goods or agriculture product supply chain management. In order to eradicate rural poverty, it is most necessary to change rural trading policy to break barriers to women's access to productive resources, and improve financial to respond to the needs of rural women producer and entrepreneurs in order to upgrade rural micro-economy [25]. So, the only way to eradicate poverty is through women economic empowerment and gender equality, especially for rural areas of developing and least developed countries. This would lead the rural women to participate in existing markets, with free access to manage production resources and to developmental projects being funded by international agencies, equality in rural trading and right to take decision for family welfare and other household activities. So, it is high time to empower rural women in economy and close gender gaps at global level to fulfil the commitments on Sustainable Development Goals, especially Goal 1 on eradicating poverty, Goal 2 on reliable security provision, Goal 3 on right to primary healthcare benefits, Goal 5 to achieve gender equality, Goal 8 to empower youth with money return working package and Goal 10 to bring equality in social inclusion activities, by 2030.

In developing countries, rural women contribute about 43% of the agricultural labour force. If rural women would be given the same opportunity to work in agriculture sector, the agricultural productivity will increase in their respective farm by 20–30%. This may be ultimately helpful in reducing the number of hungry people in the world by 12–17% [26]. Access to own land is the foremost important factor both for rural men and women. This would be the only primary solution to support agricultural production and provide food, nutrition and income security. About 5% rural women are agricultural land holders in North Africa and West Asia, while across sub-Saharan Africa, women average 15% of agricultural land holders. In addition, as compared to male-headed householders, female-headed householders have poor access to critical productivity factors like fertiliser, livestock, agricultural machineries and primary education. This sort of social trend is most often prevailing in rural areas of developing countries like Africa, Asia and Latin America [27] when more income is put into the hands of women [28–30].

By increasing the number of working women and income equality policies, the efficacy in productivity will be increased, which would be ultimately helpful in nation's economy upgradation [28]. For example, OECD (Organisation for Economic Co-operation and Development) nation like Sweden by increasing female employment rate could develop GDP by over USD 6 trillion [29]. This was a factual reflection of gender equality attitude of Sweden government [30].

Still, it has been observed that women are more likely to be unemployed than men. For example, global unemployment rates for men and women were 5.5% and 6.2%, respectively. This is noticed to be unchanged into 2018 and supposed to be the same steady state through 2021 [31].

5.3 Life Cycle of Rural Women

A woman faces five reproductive stages in reproductive life cycle. The sequential biological stages include premenarche (before the first menstrual period) stage; the reproductive, premenopausal stage; the early menopausal transition stage; the late menopausal transition stage; and, finally, menopause. So far, there is no single test to determine reproductive potential, but the menstrual cycle pattern and hormone level can give rough estimates. During such biological growth and developmental stage, a woman should avail good nutritional diet and primary healthcare facilities in order to ensure healthier motherhood for a child as potential citizen. But rural women in developing countries and least developing countries are deprived of any facilities that would be helpful during their various reproductive stages of development [31].

So, reproductive healthcare has been recognised as part of human right in the name of reproductive right [32]. Rural women are quarter of the global population (https://apolitical.co) and confront a variety of discrimination and desperate actions [32]. So it is necessary for good governance to frame suitable policies for better reproductive health services [33].

Recent report says about 830 women die from pregnancy or childbirth-related complications a day, globally. In 2015, the death cases due to pregnancy and childbirth were about 303,000. These unacceptable deaths occurred in low-quality healthcare facilities [34]. Millennium Development Goal 5 targets 75% reduction in the maternal mortality ratio (MMR). However, according to the report for the year 1990 and 2015, the global maternal mortality ratio declined by only 2.3% a year [34].

5.3.1 Infancy

In developing countries and least developing countries, in vivo growth and development of foetus mainly depend on how a pregnant lady is being nourished with nutrition-rich diet chart a day. Lack of proper diet and hygienic condition leads to high rates of stunting in children, less than 5 years of age [35]. The problem of stunting growth is associated with many other problems which include increased risk of morbidity and mortality, reduced cognitive development and short adulthood stature [36–40].

5.3.1.1 Infant Mortality

Infant mortality is defined as death of a young baby under the age of 1 year. This sort of mishappening in a community is expressed as infant mortality rate (IMR) which means number of death under 1 year per 1000 live births. The case of death of under-five mortality is known as *child mortality rate* [41, 42].

Especially, the rural women in developing countries and least developing countries face premature birth which leads to IMR [42]. In addition to this, other causes of infant mortality include pneumonia, congenital malformation, birth asphyxia, umbilical cord prolapse, neonatal infection, diarrhoea, malaria, measles and malnutrition [43].

Smoking habit of the mother is one of the potential problems for infant mortality. This could be prevented by avoiding smoking, at least in pregnancy stage [44]. Mother's level of education, environmental conditions and medical infrastructure are also responsible for infant mortality [45]. In addition, women of rural areas having poor literacy background should be well acquainted with the biological significance of sanitation, access to clean drinking water, immunisation against infectious disease and other public health measures. About three decades back, in 1990, infant (below 1 year) mortality was about 9 million. But this figure has been reduced to 4.6 million infant deaths in 2015 (Fig. 5.2).

This means infant death has declined from 65 deaths per 1000 live births to 29 deaths per 1, 000 live births [46, 47]. Globally, 5.4 million children die each year before their fifth birthday in 2017. This figure can be comparable with the horrible children cases about 12.6 million during 2013 [46]. Goal 5 of Millennium Development Goals campaign was to minimise the child mortality by 2015. The SDG programme as stated in Goal 3 also targeted to reduce child mortality to a significant number [47–49].

The risk of a child dying before completing the 1st year of age was highest in the African regions, that is, 51% per 1000 live births; as compared to the European region, this is six times higher (8% per 1000 live births) [50].

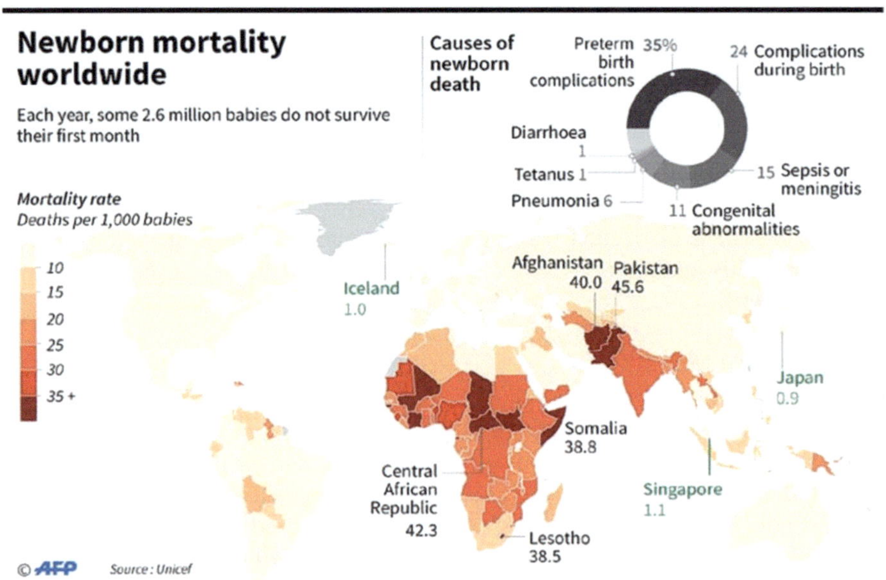

Fig. 5.2 Mortality rate by country and the main causes of newborn deaths. (With courtesy from UNICEF)

5.3.2 Puberty for Rural Women

Puberty is an earlier stage of a girl or boy just before entering the adolescence in the life cycle. It is a process of biological changes in the body of a girl between ages 10 and 14. At this stage of development, a girl undergoes both physiological and physical changes which will have impact on her general behavioural pattern [51].

However, age is a rough marker of adolescence but does not precisely define adolescence [52, 53]. Adolescence is a cumulative result of a critical stage of life influenced by social structure, environmental factors and state of psychology, education background and anthropology. Thus, the overall conclusive remark for adolescence is the transitional period between childhood and adulthood [52]. It is an orientation stage of life involving education, training and employment and a transition stage from one living circumstance to another [54]. This transition period varies by geological location and inheritance habit of a specific region.

Even within a state, adolescence age varies on the basis of social culture. Once a girl is socially accepted as mature, privileges and responsibilities like driving a vehicle, having legal sexual relation, serving in armed forces, purchasing or drinking alcohol, voting right, entering into contracts, completing certain level of education, marriage and accountability for upholding the law are allowed to be availed.

The adolescence stage of rural girls is entirely different from the social accessibility of urbanised areas of a developing countries. So, SDG programme in rural areas needs special attention. According to SDGs, by 2030 all girls and boys complete free, equitable and quality primary and secondary education leading to relevant and Goal 4 effective learning outcome. In addition, it has been ensured that all girls and boys have access to quality early childcare and pre-primary education in order to make them ready for primary education.

Adolescence is the most critical time in the second decade of life cycle while in transition to adulthood. After the completion of adolescence, between the ages of 10 and 19, a girl proceeds to womanhood, which mainly depends on how and under what circumstances a girl has experienced in her adolescence.

On the basis of rural areas location and prevailing socio-economic condition, social exclusion due to some cultural inheritance practice, poor education facilities, malnutrition like chronic problem, domestic burden and violence girls in their adolescence stage being harassed to have proper social right and justice. Despite the application of MDG and SDG programmes, as many as 48.1% girls are deprived of primary education. However, the primary school net enrolment rate in the developing regions has reached 91% in 2015, up from 83% in 2000. Sub-Saharan Africa has had the best record of improvement in primary education of any region since the MDGs were implemented [55].

Child labour is also one of the serious problems which impact the adolescence of rural girls in preventing social inclusion in a rural village life. The prevalence of child labour is highest in sub-Saharan Africa. In the least developing countries, around one in four children (ages 5–17) is engaged in labour that is considered detrimental to the health and development [56].

Menstrual hygiene management (MHM), especially in rural girls and women, is lacking due to ignorance of primary healthcare authority of the state. At least 500 million women and girls globally lack adequate facilities for MHM. Even at school level due to lack of provisions for washing water, sanitation and hygiene, the girls face challenges in maintaining their menstrual hygiene in educational institute in a dignified manner. This is one of the major reasons why rural school girls remain absent in the school for a period of 4–5 days a month [57, 58].

5.3.2.1 Sexual and Reproductive Health and Right of Rural Women

5.3.2.1.1 Reproductive Health

According to Cairo International Conference for Population and Development (ICPD) held on 1999 "Reproductive health is a state of complete physical, mental and social well-being and not merely the absence of disease or infirmity, in all matters relating to the reproductive system and to its functions and process. Reproductive health therefore implies that people are able to have a satisfying and safe sex life and how often to do so. It also includes sexual health, the purpose of which is the enhancement of life and personal relations." Thus, reproductive health means the reproductive process, functions and system at all stages of life [59]. UN agencies claim sexual and reproductive health includes physical, as well as psychological well-being vis-a-vis sexuality [60].

Sexual and reproductive health and right (SRHR) is related to multiple human rights, especially for women including right to life, free from mental and physical torture, right to healthcare services, the right to privacy, right to education and right for equality in social inclusion. The Committee on Economic, Social and Cultural Rights and the Committee on the Elimination of Discrimination against Women (CEDAW) have also mentioned the legal provision of sexual and reproductive health.

Reproductive health covers the life of women starting from conception to birth to old age passing through adolescence with sound health, excluding the period of illness and treatment. Reproductive healthcare indicators include psychotherapy; primary healthcare education; developing awareness for family planning; healthy motherhood including antenatal care, safe pre- and postdelivery care, breast feeding and quality diet with the supplement of micronutrients and essential elements; gynaecological care (prevention of abortion, treatment of complications of abortion and legalised termination of pregnancy); awareness on sexually transmitted diseases (HIV/AIDS); proper use of condom; and primary knowledge on transmission of blood-borne infections. Thus, the overall conclusion on reproductive health is to have control over sexual violence, social boycott of unusual traditional practices like female genital mutilation and popularisation of reproductive health programmes for specific group of girls under adolescence.

But violations of women's sexual and reproductive health right, especially in rural areas of developing countries and least developing countries, are noticed to be very often. This is mainly due to lack of attention of a good governance in implementing action plan, timely, for unsecure life of rural girls and women. So, unfortunately, interference in women's reproductive and sexual health, including forced sterilization, forced virginity examinations, and forced abortion without their consent is common in rural community. Due to lack of education and irrelevant social practices, women's sexual and reproductive health rights are also at risk when they are subjected to female genital mutilation (FGM) and early marriage.

In developing countries, adolescence reproductive health management is more complicated than adult reproductive health. Early pregnancy and parenting issues in puberty stage of girls create havoc, while accessing contraception. Besides this, unsafe abortions, lack of healthcare access, and high rate of HIV and sexually transmitted infection, and mental health issues are still prevailing in rural community. In addition, social custom and culture and political interference also have impact on adolescence reproductive health [61]. In adolescence, girls have yet to complete their body growth and development; therefore having pregnancy at this stage leads to complications like anaemia, psychological imbalance, possibility of HIV/AIDS infection and mental health disorders such as depression and tendency for suicidal thoughts or attempts [62]. Most of the rural teenage girls are victimised by such unfortunate mishappenings. In 2016, adolescent birth incidents between the ages of 15 and 19 were 45% per 1000 live births [63]. The top three leading causes of death in females between the ages of 15 and 19 are maternal conditions (10.1%), self-harm (9.6%) and road conditions (6.1%) [64].

Mostly, in the rural areas of developing countries, teenage pregnancy is seen often. Adolescent girls are forced to marry for dowry custom, prevailing in village communities. Besides this, due to child marriage, teenage girls feel helpless other than obliging parents. Ultimately, due to these reasons, the teenage village girls feel like a fish out of water and victimised by financial needs of girls' family, village customs, cultural norms, religious beliefs and external conflicts [65]. So, adolescent pregnancy, especially in the rural parts of developing countries, is responsible for increased health risks and intensifies poverty [66].

Social custom of forcing teenage girls into early marriage is often noticed in many parts of developing countries and least developing countries. This sort of unethical practice harms the reproductive health of adolescent girls. The World Health Organization reports [67]: "The sexual and reproductive health of the female in a child marriage is likely to be jeopardized, as these young girls are often forced into sexual intercourse with an older male spouse with more sexual experience. The female spouse often lacks the status and the knowledge to negotiate for safe sex and contraceptive practices, increasing the risk of acquiring HIV or other sexually transmitted infections, as well as the probability of pregnancy at an early age."

Niger is having the highest world record of child marriage of girls under 18, and Bangladesh has the highest rate of child marriage under age 15 [68].

The United Nations General Assembly on 19 December 2011 declared 11 October as the "International Day of the Girl Child." The day promotes girls' human rights against gender equalities and various forms of discrimination and abuse suffered by girls around the globe.

5.3.2.2 Female Genital Mutilation

One of the most horrible practices in developing countries and least developing countries is female genital mutilation (FGM) or female genital circumcision (Fig. 5.3). It is a non-medical practice. On the basis of bios, traditional beliefs and social custom, some external part of the female reproductive organ or all parts of the external genitalia are injured by the elderly members of family, without any surgical background [69]. It is mostly practised in 30 countries in Africa, the Middle East and some parts of Asia. About 200 million women and girls are victimised by this awful social practice worldwide. More severe forms of FGM are highly localised in Djibouti, Eritrea, Ethiopia, Somalia and Sudan [70].

FGM is a traditional and blind belief custom being celebrated with the leadership of an elderly family or community leader. On the basis of countries' geographical location and cultural practices, the age limit is varied. It is mostly performed on prepubescent girls, although it is belied as age ritual of girls having the intention of preserving a woman's virginity and showing loyalty to the husband after marriage [71].

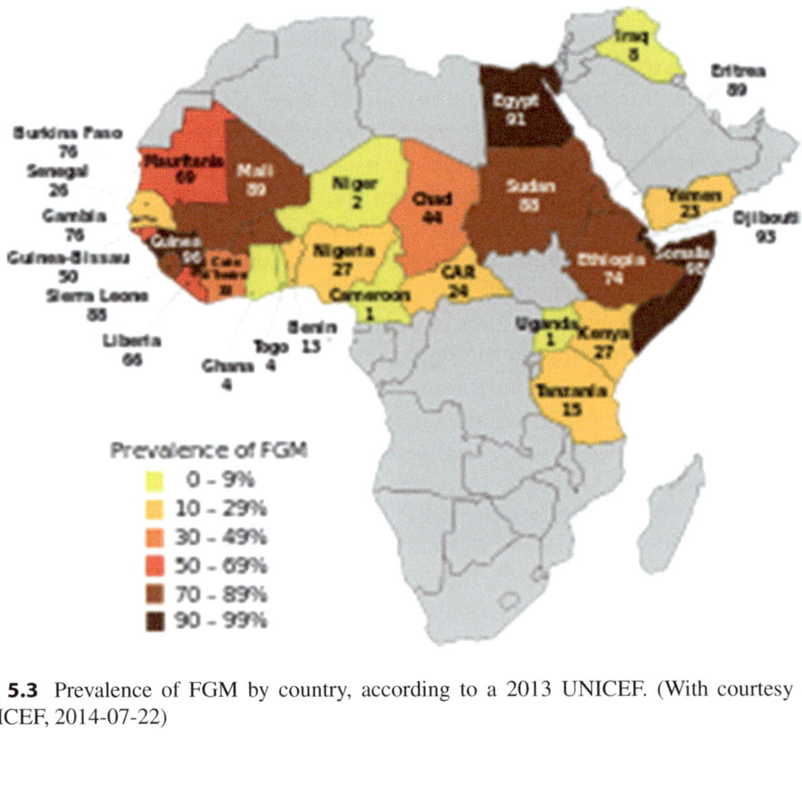

Fig. 5.3 Prevalence of FGM by country, according to a 2013 UNICEF. (With courtesy from UNICEF, 2014-07-22)

5.3.3 Maternal Health

Due to medical negligence and lack of infrastructure on maternity healthcare during predelivery period and postdelivery care, maximum maternal death has been recorded in low- and middle-income countries. It is about 100 times higher than the maternal death occurring in high-income countries, according to data from 12 million pregnancies. It was estimated that in 2015, about 303,000 women died during and following pregnancy and childbirth. Most of the cases are due to caesarean childbirth [72]. Almost all of these deaths occurred in low-resource settings, and most could have been prevented [34].

From 1990 to 2015, the global maternal mortality ratio declined by 44%, i.e. from 385 deaths per 100,000 live births (Fig. 5.4). It means the annual rate of reduction was of 2.3%. This achievement was remarkable because the target of MDG-5 was to reach the Fig. 5.5% by 2015. Mainly, two regions sub-Saharan Africa and South Asia were having about 88% of maternal deaths worldwide. In sub-Saharan Africans, maternal mortality ratio is 546 maternal deaths per 100,000 live births or 201,000 maternal deaths a year, and this is two thirds (66%) of all maternal deaths per year worldwide. The maternal deaths in South Asia were having maternal mortality ratio of 182 or 66,000 maternal deaths a year [73].

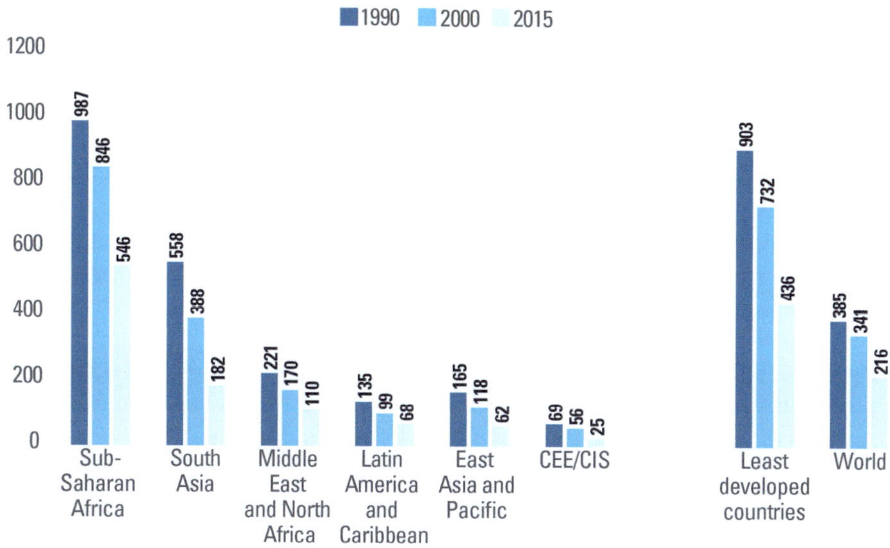

Fig. 5.4 Maternal mortality fell by almost half between 1990 and 2015. (Source: World Health Organization, UNICEF, United Nations Population Fund and The World Bank, Trends in Maternal Mortality: 1990–2015, WHO, Geneva, 2015)

Fig. 5.5 Rural women in India working in agriculture for crop plantation

5.4 Rural Health Disparities

At global level, there are significant health disparities prevailing in rural areas of developing countries, as compared to the urbanised population access to healthcare facilities and infrastructure availability. The most surprising is how, even, rural areas of developed countries like America suffer from inadequate healthcare resources and primary healthcare facilities, often characterised by indicators such as higher incidence of disease and disability, increased mortality rates, lower life expectancies and higher rates of pain and suffering. The main causes of rural risk factors are geographical isolation, poor socio-economic status, inadequate opportunity for jobs to younger generation and limited access to healthcare specialists. These sorts of discrepancies for rural people have been intensifying with the increase of population and ignorance of governance towards poor healthcare services and management for rural inhabitants. However, the SDG target is to minimise such disparities for rural people by the end of 2030.

Still, a lot of efforts have been practised to investigate and find out the existing disparities prevailing in rural areas. The Centres for Disease Control and Prevention's (CDC) is in the process of investing rural mortality and preventable deaths, health-related behaviours, chronic disease, mental health services, and related issues prevailing in rural community. In addition data bank is available on update information about rural health disparities from the Rural Health Reform Policy Research Center's (RHRPRC) publication, 2014 Update of the Rural-Urban Chartbook, and 2016 report, *Exploring Rural and Urban Mortality Differences*.

5.4.1 Obstetric and Reproductive Health Outcomes

Obstetric and reproductive disparities exist in many areas of healthcare with variable magnitude of differentiation in different countries. Even in developed nation like America, about 61 million females (31% of US female population) are members of an ethnic minority group or both [1]. It has been apprehended that non-white individuals will represent most of the US population by 2050 [74].

The National Institutes of Health defines "a health disparity population as one in which there is a significant disparity in the specific population as compared with the general population" [75]. Mostly, disparities in health and healthcare are noticed among women who are members of racial and ethnic minority groups. Generally, obstetrician-gynaecologists face this problem while distinguishing between disparities in health conditions and outcomes (e.g. presence of uterine fibroids or early prenatal care or mammography). Women community from, both, ethnic minority group and white individual require different solutions, although each category has resources to improve their understanding of the aetiologies of such disparities and engage in strategies to minimise them.

Health disparities, especially in rural areas of developing countries, are mainly due to low income, lack of healthcare infrastructure, illiteracy and unemployment problem for youth. Meanwhile, after the implementation of SDG drive, reduction rates in health disparities are reduced.

Mostly, racial and ethnic disparities in healthcare are due to socio-economic condition, microenvironment and social customs and culture rather than any biological factors. Often, the frequency of certain genetic variations may be confused with racial or ethnic groups but generally related to a common ancestral lineage. For example, sickle cell disease is an increased frequency in individuals of African descent. Genetic polymorphisms associated with increased susceptibility to disease also may vary in frequency in different racial and ethnic groups [76].

Sometimes, racial and ethnic disparities in healthcare may be due to variation in patient psychology and attitudes but not observed differences. For example, women belonging to ethnic minorities may not prefer to have prenatal genetic screening, which can give important clue to understand the basic cause of a disease [8]. Physical activities and dietary practices may vary by race and ethnicity and, therefore, contribute to differences [77].

Variation in patient preferences, attitudes and adherence to treatment plans may explain some, but certainly not all, observed differences. For example, racial and ethnic minorities may be less likely to undergo prenatal genetic screening, but this difference is partially explained by differences in value placed on the information obtained from such testing [77]. Studies suggest the likely effect of experiences of racism and life stressors on obstetric and gynaecologic outcomes [78]. Behaviours that affect health status, such as physical activity and dietary practices, may vary by race and ethnicity and, therefore, contribute to differences [79].

In the United States, healthcare system is taken care of by market forces rather than a responsibility of the government. Thus, the healthcare system in the United States is not accessible to poor citizens who do not have capacity for optional health insurance. As a result rural women belonging to minority having low socio-economic status are adversely affected by the availability of healthcare services [80, 81]. Varying geographic availabilities of healthcare institutions also may contribute to racial and ethnic disparities in healthcare [82, 83].

5.4.2 Access to Health Services

Access to health services means "the timely use of personal health services to achieve the best health outcome" [84]. It is having three distinguished steps: (i) entry to healthcare system through insurance coverage, (ii) access to all places of different geographical locations and (iii) identifying a trusty healthcare councillor through whom the patients will get safe and reliable services. All the people of the world have the right to access minimum primary healthcare services. Primary healthcare is targeted for health and well-being of the people on an as-needed basis, irrespective of any geographical location of a country. It addresses inclusive and unified physical, mental and social health and welfare.

Primary healthcare services focus on the entire healthcare of an individual, not just for little specific disease. Primary healthcare assures a comprehensive package of healthcare, ranging from promotion and prevention to treatment, rehabilitation and smoothing care to fulfil the needs of the day for a comfortable, healthy family life [85].

The WHO's definition of primary healthcare (PHC) refers to "essential healthcare" on the basis of scientific conceptual practices and socially acceptable clinical healthcare technology which can be helpful in accessing universal healthcare practices for all the people of the world at community, family and individual level. PHC is to take care of health beyond the traditional social practices without the basic concept on the criteria of health management [86]. PHC has broad spectrum of health coverage which includes access to health services, surrounding microenvironment and lifestyle [87]. So, PHC and public health measures in combination develop the basic theme of universal health system [88]. The World Health Organization defines three major components of PHC as "empowering people and communities, multisectoral policy and action; and primary care and essential public health functions as the core of integrated health services [84]." So, based on the aforesaid facts, PHC is to take care of an individual starting from primary clinical diagnosis to preventive measures, after gaining through information on an individual health. PHC is the most efficient and cost-effective way to achieve universal health coverage globally.

However, the main barriers to health services include (i) high cost of healthcare, (ii) lacking insurance coverage and (iii) lacking of social customs and culture competent care.

5.4.3 Obstetric-Gynaecologic Workforce in Rural Areas

At present, less than about 50% of the world's people live in rural areas. This figure is going to decrease, due to the migration of rural people to urbanised areas, and it is expected that by 2050, the urbanised population will reach about 68%. It is apprehended that gradual shift in residence of the human population from rural to urban areas, combined with the overall growth of the world population, uphold another 2.5 billion people to urban areas by 2050. The most victimised nations for such population imbalance are Asia and Africa. According to the report from Population Division of the UN Department of Economic and Social Affairs (UN DESA), the intensification of urban population will be more likely to be in developing countries like India and China and in least developing country like Nigeria, by the end of 2050. The urban population has reached from 751 million in 1950 to 4.2 billion in 2018. In spite of relatively slow urbanisation in Asian countries, at present about 50% urbanisation is prevailing due to slow migration of rural population. In contrast, Africa remains mostly rural, with 43% of its population living in urban areas. Presently, most urbanised regions include Northern America (with 82% of its population living in urban areas in 2018) followed by Latin America and the Caribbean (81%), Europe (74%) and Oceania (68%) [89].

Most of the rural areas are geographically isolated from urbanised locality, mainly due to the lack of suitable logistic systems. In addition most health workers stay and work in cities. This imbalance in health workforces is noticed to be common in most of the developing countries. The most acute sufferers for such deficiency in health workers are the people of low-income countries.

In order to fulfil the Sustainable Development Goals, we need huge workforce of over 18 million additional health workers, all over the world by 2030. The deficiency in the availability of healthcare workers is mainly in low- and lower-middle-class-income countries. The growing demand for health workers is supposed to create 40 million health sector jobs to the global economy by 2030 [89]. Besides this, UHC emphasises not only the wide ranges of service areas but also the ways and means of funding management and delivery process. In addition UHC manages the integrated healthcare services by focusing on the needs of people and communities. Healthcare services include traditional and complementary medicine services based on the need of the people and communities by giving moral support to intensify their active participation in their health and health system.

So, 12 December is celebrated as Universal Health Coverage Day, all over the world. It is promoted by the WHO. UHC plays an important means to bring awareness among the people on "right to health," based on international mutual understanding being supported by a number of international agreements or widely affirmed international documents. UHG coverage has been included in the new SDGs for 2015–2030 adapted by the United Nations. Most of the developing countries suffer from inclusive healthcare due to the extreme negligence of governance in making proper plan and framing suitable policies for rural healthcare services. In 2015, the WHO and UNICEF reported that 38% of healthcare facilities surveyed did not have access to a basic water source and 35% lacked the materials necessary for people to wash their hands effectively. For quality health services, water, sanitation, hygiene, waste management and environmental cleaning (WASH) services in healthcare facilities are most essential. Such services are not only beneficial to patients but also reduce the burden of infection, input of hazardous biological waste to the environment and risk to the community. With the lack of proper WASH services in most of the healthcare centres, the WHO, in collaboration with UNICEF and with internationally recognised voluntary organisation, is in a process of upgrading WASH in healthcare worldwide. In order to increase the efficacy of these services, the WHO and UNICEF have aligned WASH, directly with SDGs (Goals 3.1, 3.2, 3.8 and Goal 6) to reduce unfortunate incidents like maternal mortality, ending preventable newborn deaths and developing access for quality universal health coverage [90, 91].

The international recognised voluntary organisations working as driving forces for the success of WASH programme include The Rockefeller Foundation, World Bank Group, World Health Organization, Oxfam, Save the Children, Management Sciences for Health, Action for Global Health, Global Parliamentary Network on UHC and Public Health Foundation of India. Forty-two academic institutions and about 792 other nongovernmental organisations (NGOs) are involved as global partners [92].

So, it should be mandatory to all developing countries to frame proper rural healthcare service policy giving special emphasis to provide enough healthcare workers, irrespective of their level of economic development and struggle to achieve health equality and to meet the health needs of their population, especially vulnerable and disadvantaged groups. The governance policy should emphasis the government to provide trained healthcare workforce as and when an individual needs at rural community level. In addition the governance of the state should frame suitable policies which could able to regulate the movement of health workers within the boundaries of a country and develop new strategies to ensure the availability of health workers in remote and rural areas by providing improved and attractive service benefits and facilities for their respective family education and quick transport system for their movement to urbanised areas. The governance should recommend the government the current working policy of WHO on the basis of Global Code of Practice on the International Recruitment of Health Personnel [93].

In order to have control over the incidents of migration of health workers, a worldwide phenomenon, the WHO Global Code of Practice on the International Recruitment of Health Personnel was adapted by the 63rd World Assembly on 21 May 2010.

The basic aim of the Code is to establish and promote voluntary principles and practices for the ethical international recruitment of health personnel and to facilitate the strengthening of health systems. The Code was designed by member states to serve as a continuous and dynamic framework for global dialogue and cooperation.

5.4.4 Rural Women's Health Services

Position of women in rural society is most significant. Women play a key role in achieving food and nutrition security, family management, generating income and improving rural livelihood and overall welfare of the family and community. Rural women are part and partial of rural agriculture development and home industry management. In addition rural women are the main instruments in the implementation and drive for Sustainable Development Goals adopted in 2015. In spite of so much significant contribution, rural women are confronted with tough health services. This includes limitation on their movement, lack of access to transport and social exclusion. Besides this, rural women are deprived of holistic health services to support their physical as well as mental and emotional well-being.

The life expectancy gap between men and women is about 4.6 years globally. Generally, women life tenure is about 10 years more than men [94]. Besides this, the global burden of disease exceptionally affects men's life tenure as compared to women [95], although women are more likely to spend longer time living with a disability [96].

For example, in most of the rural locality, both men and women are addicted to tobacco. But women are noticed to be earlier victimised than men. Women appear to develop severe chronic obstructive pulmonary disease at a younger age than men [97].

According to the WHO, gender-based social behavioural attitudes are responsible for health vulnerabilities and risks of rural women and girls, which are identified and addressed in health service delivery. Gender discrimination is not only a factor for social exclusion but also, directly or indirectly, acts as a negative factor, responsible for depriving rural women from primary healthcare opportunities. So, SDG 3 is targeted to ensure, irrespective of gender discrimination, healthy lives and promote well-being for all at all ages. It aims by 2030 to reduce global maternal mortality ratio to less than 70% per 100,000 live births; to end preventable deaths of newborns and children under 5 years of age; to end the epidemics of AIDS, tuberculosis, malaria and neglected tropical diseases and combat hepatitis, water-borne disease and other communicable diseases; to reduce by one third premature mortality from non-communicable diseases; to prevent drug abuse and alcohol intake, especially in women and young girls; to ensure universal access to sexual and reproductive healthcare services; to encourage all developing countries to frame policies for universal health coverage; and to reduce accidental death rates due to hazardous chemicals, gas and contaminated water.

Thus, SDG Goal 5 is targeted to get rid of all forms of social exclusion and violence presently prevailing in most of the rural community of developing countries and least developing nations. In addition Goal 5 gives emphasis on equal rights of rural women to economic resources and access to ownership of property.

Both SDG 3 and 5 have crosscutting features for gender equality. In SDG 3, there are six gender-specific indicators which include (i) maternal mortality ratio, (ii) birth attended by skilled health personnel, (iii) HIV infection by sex, (iv) modernised family planning methods, (v) adolescent birth rate and (vi) coverage of essential health services including reproductive and maternal health.

Besides the adaptation of SDG programmes, there are other agencies also involved in overall well-being for rural women through advocating for social inclusion, job opportunity for rural women with satisfactory rumination package and complete restriction of child labour.

5.4.5 Initiatives to Improve Services for Rural Women

The Committee on Economic, Social and Cultural Rights, UN, in its 22nd session, held in Geneva, 2000, defined the right to the highest attainable of health (article12 of the International Covenant on Economic, Social and Cultural Rights) "Health is a fundamental human right indispensable for the exercise of other human rights. Every human being is entitled to the enjoyment of the highest attainable standard of health conducive to living life in dignity. The realization of the right to health may be pursued through numerous, complementary approaches, such as the formulation of health policies, or the implementation of health programmes developed by the World Health Organisation (WHO), or the adaptation of specific legal instruments. Moreover, the right to health includes certain components which are legally enforceable."

It implies a clear set of legal obligations on all the states of America to ensure appropriate enjoyment of health services for all people without discrimination. Right to health is an internationally recognised set of agreements, equivalent to other human rights to food, housing, work, education, information and social inclusion. The right to health, as with other rights, includes both freedoms (right to control one's health and body without any short of interference) and entitlements (to enjoy the highest attainable level of health). Each and every individual of a rural community should have access to primary healthcare, dental care, behavioural health, emergency healthcare and public health services.

The WHO has been greatly involved in accelerating the action plan on SDGs and committed to mainstream human right into healthcare programmes and policies on national and regional level worldwide. Besides this, the WHO is actively involved in providing financial and moral support to operate SDGs in a successful way to bring social amendment in rural women's lifestyle, irrespective of age and gender.

Currently, a lot of nongovernment organisation (NGOs), institutes and universities are involved in nullifying the difficulties while providing healthcare services to rural women. For example:

The University of Texas Medical Branch at Galveston has developed an off-site clinical system that offers a wide range of services to low-income women and their infants over large geographical areas. In addition, the rural women's community status development is linked to culturally relevant services and transportation needs and uses electronic medical records to facilitate continuity of care. The program addresses culturally relevant services and transportation needs and uses electronic medical records to facilitate continuity of care [98].

In 2017, to help the rural women from the fatal issues on maternal death, the American College of Obstetricians and Gynecologists (ACOG) addresses the challenges of maternal death rates in rural America, especially Guam, American Samoa, Alberta, British Columbia and Central America [99].

Wyoming is a state having poor healthcare for pregnant women or infants with limited number of paediatric specialists. Wyoming approves out-of-state healthcare providers and facilities as state Medicaid providers with the reimburse transport to care and delivery in an out-of-state subspecialty hospital as and when medically require [100].

Arkansas is a state in the southern region of the United States. This state has been confronting high poverty and suffering from inadequate infrastructure for specialised maternal-foetal medicine. According to the Arkansas Department of Health, 73 out of 75 counties of this state are having extremely poor facilities for healthcare services which are responsible for high-risk pregnancies and low-birthweight babies. In order to overcome such problems, the University of Arkansas for Medical Sciences (UAMS) implemented a state-wide telehealth programme known as The Antenatal and Neonatal Guidelines, Education and Learning System (ANGELS) which can enhance rural women's accessibility to physicians and high-risk pregnancy services at the University of Arkansas for Medical Sciences [101].

5.5 Changing the Rural Women's Life Pattern

Rural women, the key elements of rural structure and function, work for day and night for the well-being of rural community. Neither have they cared for sunshine, heavy rain or adverse climatic changes nor are busy in keeping the social integrity and welfare of their respective family and communality. For clean drinking water and firewood, they go on moving from dawn to dusk by forgetting themselves and only thinking for the family well-being. Without rural women and girls, rural communities would not survive. Still, rural women and girls are suffering from poverty, ill health and social exclusion. In order to bring awareness about the development and enhance social inclusion of rural women, 15 October is celebrated as "International Day of Rural Women" having the motto: "Sustainable infrastructure, services and social protection for gender equality and the empowerment of rural women and girls," closely followed by World Food Day (16 October) and International Day for the Eradication of Poverty (17 October). In these consecutive 3 days (15–17 October), the world celebrates for better public services, including healthcare, education, childcare and shelters, on which millions of rural women depend on and laws, policies and budgets to improve their livelihoods and well-being. Other activities and events held to promote the day include:

- International agriculture exchange programme with special reference to rural women
- Funding for rural women uplift programmes
- Initiation of training and workshops for rural women on the latest development on agriculture and healthcare

- Rural women and girl empowerment programmes with governance policymakers

5.6 Rural Women's Development Project

The main targets of projects based on rural women's lifestyle development are to strengthen process to support and promote the present lifestyle of rural women, confronting poverty, lack of healthcare facilities and social exclusion. So, the main objectives include (i) to peruse the state governance to frame policies on implementation of women's self-help groups (SHGs) so that the rural women community can able to project their problem for equality in community life, (ii) to frame effective policies to support women's empowerment projects, (iii) to create community assets for women's welfare and (iv) to provide effective project management systems.

Under women's empowerment in rural community-driven development (CDD) project [102], it is intended to empowering rural community, especially women communities that have been ignored while evaluating social or community development work in rural areas of developing countries. This could have been due to lack of gender-specific outcomes as explicit objective, and social inclusion projects are not specified regarding the participation of women. Community-driven development has potential to address the persistent gender gap that brings restriction in women's active participation in public decision-making. CDD can catalyse development and enhance women's in becoming rural women as part and partial of decision-making, instead of simple participation.

5.7 Empower Rural Women to End Hunger

Especially in rural areas, women are the key elements for developing and managing the structural and functional get-up of rural areas by supporting agricultural practices, farmhouse development, food production and rural home industries development and play an active role as caregiver and community leaders.

In developing countries, women are accountable for 45% agriculture labour force, ranging from 20% in Latin America, 60% in certain part of Africa and Asia and 85% in India. In most of the rural India, women work as agricultural and farm labourers, in addition to performing all types of household works, including taking care of children (Fig. 5.5).

Improvement of social status (social inclusion, funding for agriculture, nutritious food, gender equality) of rural women will certainly improve the agriculture production and food security. Evidence from Africa, Asia and Latin America have encouraging result in food security improvement as a result of women's access to productive resources, modern agriculture technologies, land, market opportunities, employment and social protection. For example, Guinea is one of the least developing countries. But by cooperative rural farming (with the support of UN Women), the women of Guinea started mass cultivation of moringa (*Moringa oleifera*) which opened a major source of avenue. The vitamin-rich leaves and pods of the tree are in good demand by international markets and help in prevention of soil erosion (Fig. 5.6).

This plant is native to the sub-Himalayan areas of India, Pakistan, Bangladesh and Afghanistan. It is used for anaemia, arthritis and rheumatism, asthma, prevention for cancer, constipation, diabetes, diarrhoea, epilepsy, intestinal ulcers, high blood pressure, kidney stone, fluid retention, thyroid disorders and bacterial, fungal and parasitic infection.

So, this highly adaptable wonder plant is not only helpful in improving women's social and economic status within their households and communities but has a direct impact on food security and nutrition, in particularly on child nutrition.

Rural women should be well acquainted with modern technologies for healthcare and communications. For example, a simple mobile phone can have access to know about climatic changes and weather condition for rural women involved in agriculture and marketing.

In 2016, Rwanda launched a programme "Buy from Women" by the UN Women and World Food Programme, funded by the government of China and Finland. This was beneficial to over 3000 farmers from 12 maize farming cooperative to market their products, outside their territory. Besides this, the farmers have opportunity to sign contracts with maize buyers, forging stronger links to markets.

In most of the rural areas of developing countries, women are lacking right to have ownership of land. This practice is a sort of social discrimination and even violence. The UN Women, the Food and Agriculture Organization of the United

Fig. 5.6 *Moringa oleifera* aka drumstick tree for agroforestry. (Courtesy: ICAR-CIRG, Makhdoom, Mathura (U. B. Chaudhary))

Nations and the International Labour Organisation jointly developed funding for widow for procuring cultivated land of her own. This is a milestone practice being organised by international agencies for rural women to have agricultural land ownership.

Most of the rural areas of developing countries and least developing countries are lacking of adequate electricity. Conventional form of electricity supply system faces heavy financial burden to a state government due to the uneven distance location of remote rural areas. About 1 billion rural people still do not have access to electricity. The only solution for such energy problems is to implement nonconventional energy system, accordingly the availability of natural resources of a particular geographic location. In order to implement solar energy-based electric power generation in Liberia, an African country, Musu Junius and Marie Weeks attended Barefoot College solar engineering training in India funded and supported by UN Women.

In Quito, nature and sustainability are considered as complementary of developing programme of city's conservation. For such purpose the city authority of Quito operates a project on Sustainable Agriculture with Gender Inclusion and Participation work on food security, urban agroecology, climate change adaption and nutrition. This programme is on operation condition with the cooperation of women and female-headed households for over a decade to upgrade the quality of life of the vulnerable people of the Metropolitan District of Quito.

By giving full liberty and financial equality to rural women in all types of rural services related to social uplift and community development, they will feel free to spend money more of than men do on food, clothing and education for their children. It means empowering rural women is a process of translating into overall improvement in well-being for children, households and communities, which in turn contributes to building human capital for future generation and for long-term social and economic growth.

References

1. Ransom EI, Elder LK (2013) Nutrition of women and adolescent girls: why it matters. Population Reference Bureau. Available online: http://www.prb.org/Publications/Articles/2003/NutritionofWomenandAdolescentGirlsWhyItMatters.aspx
2. FAO, IFAD and WFP (2015) The state of food insecurity in the world: meeting the 2015 international hunger targets: taking stock of uneven progress. FAO, Rome. doi:I4646E/1/05.15 Available online: http://www.fao.org/3/a-i4646e.pdf
3. UNICEF (2014) Low birth weight: current status and progress. 2014 Low birth weight statistical table, (NFHS 2005–2006), Analyzed by UNICEF HQ June 2008. The UNICEF low birth rate data website http://data.unicef.org/nutrition/low-birthweight.html
4. Low birthweight-UNICEF DATA (2015). https://data.unicef.org
5. Delisle HF (2008) Poverty: the double burden of malnutrition in mothers and the intergenerational impact. Ann N Y Acad Sci 1136(1):172–184
6. De Schutter O (2013) Gender equality and food security: women's empowerment as a tool against hunger. FAO, Manila. (2012). Gender and nutrition. http://www.fao.org/docrep/012/al184e/al184e00.pdf
7. Malnutrition-World Health Organisation (2018). https://www.who.int

8. Bailey RL, West KP Jr, Black RE (2015) The epidemiology of global micronutrient deficiencies. Ann Nutr Metab 66(2):22–33
9. Balarajan Y, Ramakrishnan U, Ozaltin E et al (2011) Anaemia in low-income and middle-income countries. Lancet 378:2123–2135
10. Stevens GA, Finucane MM, De-Regil LM (2013) Global, regional, and national trends in haemoglobin concentration and prevalence of total and severe anaemia in children and pregnant and non-pregnant women for 1995–2011. Lancet Glob Health 1(1):e16–e25
11. McLean E1, Cogswell M, Egli I (2009) Worldwide prevalence of anaemia, WHO vitamin and mineral nutrition information system, 1993–2005. Public Health Nutr 12(4):444–454
12. Rukuni R, Bhattacharya S, Murphy MF (2016) Maternal and neonatal outcomes of antenatal anaemia in a Scottish population: a retrospective cohort study. Acta Obstet Gynecol Scand 95(5):555–564
13. Young EM (2012) Food and development. Routledge, Abingdon, pp 36–38. ISBN:978-1-135-99941-4
14. Jones & Bartlett Publishers (2011) Essentials of International Health. ISBN 978-1-4496-6771-9, p 194
15. Facts for life (4th ed.) (2010) New York: United Nations Children's Fund. 2010, pp 61–75. ISBN:978-92-806-4466-1
16. Stoelting's anaesthesia and co-existing disease (6th Ed.) (2012) Philadelphia: Saunders/Elsevier, p 324. ISBN 978-1-4557-3812-0
17. The State of Food Insecurity in the World (2018) Food and Agricultural Organization of the United Nations. Retrieved January 11, 2019
18. Global hunger declining, but still unacceptably high International hunger targets difficult to reach (2010) Food and Agriculture Organization of the United Nations. September 2010. Retrieved July 1, 2014
19. GBD (2015) Mortality and causes of death, collaborators (October 8, 2016). Global, regional, and national life expectancy, all-cause mortality, and cause-specific mortality for 249 causes of death, 1980–2015: a systematic analysis for the global burden of disease study 2015. Lancet 388(10053):1459–1544
20. GBD (2013) Mortality and causes of death, collaborators (December 17, 2014). Global, regional, and national age-sex specific all-cause and cause-specific mortality for 240 causes of death, 1990–2013: a systematic analysis for the global burden of disease study 2013. Lancet 385(9963):117–171
21. World Bank (2012) Promoting women's agency. World Development Report, 150–192. Available online: https://openknowledge.worldbank.org/handle/10986/4391
22. The World Bank (2017) The World Bank released new poverty lines-Find out where your country stands, https://www.businessinsider.com
23. United Nations Children's Fund (UNICEF), World Health Organisation (2017). https://www.unicef.org/publications,index
24. FAO (2011) The State of Food and Agriculture 2010–2011 (SOFA). Rome. p 16–17
25. FAO, IFAD (2010) Gender and rural employment policy. Rural women's entrepreneurship is "good business"!
26. FAO (2011) The State of Food and Agriculture 2010-2011 (SOFA). pp. 16–17. Rome. (Repeat)
27. FAO (2011) Chapter 3. Documenting the gender gap in agriculture, pp. 23–38. www.fao.org By increasing the number of working women and income equality policy, the efficacy in productivity will be increased, and which would ultimately helpful in nation's economy up gradation (28). For example, OECD (Organization for Economic Co-operation and Development) nation like Sweden by increasing female employment rate could develop GDP by over USD 6 trillion (29). This was a factual reflection of gender equality attitude of Sweden government (30)
28. International Monetary Fund (2018) Pursuing women's economic empowerment. https://www.imf.org/en/Publications/PolicyPapers/Issues/2018/05/31/pp053118pursuing-womens-economic-empowerment

29. PwC, Women in Work Index (2018). At: https://www.pwc.co.uk/services/economics-policy/insights/women-in-work-index.html
30. Cuberes D, Teignier M (2016) Aggregate effects of gender gaps in the labour market: a quantitative estimate. J Hum Cap 10(1):1–32. https://doi.org/10.1086/683847
31. International Labour Organization (2018) World employment and social outlook: trends for women 2018: global snapshot (Geneva), http://www.ilo.org/wcmsp5/groups/public/%2D%2D-dgreports/%2D%2D-dcomm/%2D%2D-publ/documents/publication/wcms_619577.pdf
32. Zanjani H, Jazaeri A (2015) A comparative study of effective on women's reproductive health among Darood and Koohdasht residents. Journ Ilam Univer 22(6):103–114
33. WHO (2002) Report of a technical consultation on sexual health. Geneva 2002. Defining sexual health
34. Alkema L et al (2016) Global, regional, and national levels and trends in maternal mortality between 1990 and 2015, with scenario-based projections to 2030: a systematic analysis by the UN maternal mortality estimation inter-agency group. Lancet 387(10017):462–474
35. Christian P et al (2013) Risk of childhood undernutrition related to small-for-gestational age and preterm birth in low-and middle-income countries. Int J Epidemiol 42(5):1340–1355
36. Crookston BT (2013) Post infancy growth, schooling, and cognitive achievement: young lives. Am J Clin Nutr 98(6):1555–1563
37. Mason JB et al (2012) The importance of maternal undernutrition for maternal, neonatal, and child health outcomes: an editorial. Food Nutr Bull 33(2):S3–S5
38. Prendergast AJ, Humphrey JH (2014) The stunting syndrome in developing countries. Paediatr Int Child Health 34(4):250–265
39. Victoria CG et al (2001) Short-term benefits of catch-up growth for small-for-gestational-age infants. Int J Epidemiol 30(6):1325–1330
40. Walker SP (2011) Inequality in early childhood: risk and protective factors for early child development. Lancet 378(9799):1325–1338
41. Child Mortality (2017) Under-five mortality. UNICEF, https://data.unieef.org
42. Infant Mortality (2017) What is CDC doing? Infant Mortality | Maternal and Infant Health | Reproductive Health |. Centers for Disease Control and Prevention. https://www.coursehero.com
43. Infant Mortality & Newborn Health (2017) Women and children first. https://www.philips.co
44. Hall ES, Venkatesh M, Greenberg JM (2016) A population study of first and subsequent pregnancy smoking behaviours in Ohio. J Perinatol 36(11):948–953
45. Genowska A et al (2015) Environmental and socio-economic determinants of infant mortality in Poland: an ecological study. Environ Health 14:61
46. Roser M (2013) Child mortality. Our World in Data. https://ourworldindata.org
47. Mortality rate, infant (per 1,000 live births). Data (2019). data.worldbank.org. https://en.wikipedia.org
48. WHO | Child mortality (2017). www.who.int
49. Sustainable development knowledge platform". UN-DESA. https://sustainabledevelopment.un.org
50. World Mortality Organization (2017) Global Health Observatory (GHO) data, https://www.who.in
51. Kail RV, Cavanaugh JC (2010) Human development: a lifespan view (5th ed.). Cengage Learning. ISBN 978-0-495-60037-4, p. 296.
52. Larson R, Wilson S (2004) Adolescence across place and time: globalization and the changing pathways to adulthood. In: Lerner R, Steinberg L (eds) Handbook of adolescent psychology. Wiley, New York
53. Žukauskaitė S et al (2005). Lašienė D, Lašas L, Urbonaitė B, Hindmarsh P (September 2005)) Onset of breast and pubic hair development in 1231 preadolescent Lithuanian schoolgirls. Arch Dis Child 90(9):932–936
54. Arnett JJ (2007) Emerging adulthood: what is it, and what is it good for? Child Dev Perspect 1(2):68–73

55. The Millennium Development Goals Report (2015), United Nations, https://www.un.org
56. UNICEF (2013) Impact of unpaid household services on the measurement of child labour. MICS Methodological UNICEF, New York, P-2
57. Menstrual Hygiene Management (2015). https://www.unicef.org
58. Aditi V et al (2018) School absenteeism during menstruation amongst adolescent girls in Delhi, India. J Fam Community Med 25(3):163–168
59. WHO: Reproductive health (2017). https://www.who.int
60. International technical guidance on sexuality education: an evidence-informed approach. Paris: UNESCO. 2018. p. 22. ISBN 978-92-3-100259-5
61. Morris JL, Rushwan H (2015) Adolescent sexual and reproductive health: the global challenges. Int J Gynaecol Obstet 131(1):S40–S42
62. World Health Organization (2018) Maternal, newborn, child and adolescent health, https://www.who.int
63. United Nations Population Division. Adolescent fertility rate (births per 1,000 women ages 15–19). https://www.inationmaster.com
64. Panduan Teknis SIJARI EMAS. USAID. WHO (2017). Global strategy for women's, children's and adolescents' health (2016–2030) key statistics
65. An R, Jackson E, Dunham S (2018) Girl child marriage: a persistent global women's health and human rights violation. Global perspectives on women's sexual and reproductive health across the life course. Springer, Cham, pp 3–19
66. World Health Organization (2011) Preventing early pregnancy and poor reproductive outcomes among adolescents in developing countries. https://www.who.int
67. World Health Organisation (2012) Child marriage – a threat to health. www.euro.who.int
68. International Conference on Population and Development (1994). https://www.unicef.org
69. World Health Organisation (2018) Female genital mutilation https://www.who.int.
70. Sexual and reproductive health (2018) Female genital mutilation (FGM), https://www.who.int
71. Klein E et al (2018) Female genital mutilation: health consequences and complications-a short literature review. Obstet Gynaecol Int 2018:7365715
72. World Health Organisation (2019) Maternal healthy. https://www.who.int/maternal-health/en
73. UNICEF for every child (2017) Maternal mortality. https:// data.unicef.org
74. Taylor P, Cohn D (2012) A milestone en route to a majority minority nation. Pew Research Center, Washington, DC; http://www.pewsocialtrends.org/2012/11/07/a-milestone-en-route-to-a-majority-minority-nation/#
75. National Institutes of Health (2013) NIH health disparities strategic plan and budget fiscal years 2009.2013. http://www.nimhd.nih.gov/documents/NIH%20Health%20Disparities%20Strategic%20Plan%20and%20Budget%202009-2013
76. Romero R et al (2002) The design, execution, and interpretation of genetic association studies to decipher complex diseases. Am J Obstet Gynecol 187:1299–1312
77. Kuppermann M et al (2006) Beyond race or ethnicity and socioeconomic status: predictors of prenatal testing for Down syndrome. Obstet Gynecol 107:1087–1097
78. Rosenthal L, Lobel M (2011) Explaining racial disparities in adverse birth outcomes: unique sources of stress for Black American women. Soc Sci Med 72:977–983
79. Clark A, Fong C, Romans M (2002) Health disparities among U.S. women of colour: an overview. Jacobs Institute of Women's Health, Washington, DC. http://www.jiwh.org/attachments/Health%20Disparities%20Overview
80. Health disparities in rural women (2014) Committee Opinion No. 586. Am Coll Obstetricians Gynaecologists Obstet Gynecol 123:384–388
81. Health cares for urban American Indian and Alaska Native women (2012) Committee Opinion No. 515. Am Coll Obstetricians Gynaecologists Obstet Gynecol 119:201–205
82. Henry JK (2013) Family Foundation. Health coverage for the black population today and under the affordable care act. KFF, Menlo Park; http://kff.org/disparities-policy/fact-sheet/health-coverage-for-the-black-population-today-and-under-the-affordable-care-act

83. Health cares for unauthorized immigrants (2015) Committee Opinion No. 627. Am Coll Obstetricians Gynaecologists Obstet Gynecol 125:755–759
84. World Health Organization (1978) Declaration of Alma-Ata. Adopted at the International Conference on Primary Health Care, Alma-Ata, USSR
85. Starfield B (2011) Politics, primary healthcare and health. J Epidemiol Community Health 65:653–655
86. Public Health Agency of Canada (2011) About Primary Health Care.
87. Marcos C (2004) The Origins of primary health care and selective primary health care. Am J Public Health 22:1864–1874
88. White F (2015) Primary health care and public health: foundations of universal health systems. Med Princ Pract 24:103–116
89. United Nations (2018) Revision of world urbanization prospects, Department of Economic and Social Affairs, https://www.un.org/development/.../2018-revision-of-world-urbanization-prospects.ht
90. WHO, United Nations' Children's Fund (2015) Water, sanitation and hygiene in health care facilities – WHO (2015); Status in low- and middle-income countries and way forward, https://www.who.int/water sanitation health
91. Health care facilities –WHO (2016). http://www.who.int/water_sanitation_health/Facilities/en/
92. Health for All (2017) Global Partners of UHC Day, https://www.healthforall.org/partners/
93. World Health Organisation (2015) Managing health workforce migration – the Global Code of Practice. https://www.who.int/hrh/migration/code/practice/en
94. Life expectancy (2018) Global Health Observatory (GHO) data [internet]. Available from: http://www.who.int/gho/mortality_burden_disease/life_tables/situation_trends_text/en/
95. GBD Compare (2016) Data visualization hub [internet]. Institute for Health Metrics and Evaluation, University of Washington, Seattle, http://vizhub.healthdata.org/gbd-compare
96. Gender and blindness (2002) Gender and health information sheet [internet]. World Health Organization, Geneva. Available from: http://www.who.int/gender-equity-rights/knowledge/a85574/en
97. The health consequences of smoking (2014) 50 years of progress. A report of the Surgeon General. Centers for Disease Control and Prevention, Atlanta; https://www.surgeongeneral.gov/library/reports/50-years-of-progress/full-report
98. Planning, protection, prevention: reducing unintended pregnancy in Colorado (2009) Denver (CO): PFC; 2009. p. 45–51. http://www.preventionfirstcolorado.org/uploads/PPP-Rec.7.AdvancedPracticesNurses
99. Sharon TP, Linda MW (2018) Maternal death in rural America, obstetrics- gynaecology and women's health. Matern Mortal Spec Rep 64:08
100. Wyoming Department of Health. State narrative for Wyoming: application for 2014 annual report for 2012 (2013) Maternal and child health services title V block grant. Wyoming Department of Health, Cheyenne; 2013. http://www.health.wyo.gov/Media.aspx?mediaId=13622. Retrieved October 30, 2013
101. Lowery C et al (2007) ANGELS and University of Arkansas for Medical Sciences paradigm for distant obstetrical care delivery. Am J Obstet Gynecol 196:534.e1–534.e9
102. World Bank (2016) Women's empowerment in rural community-driven development projects. https://www.findevvgateway.org

Sustainable Livestock Farming for Zero Hunger

6

Sustainability in livestock farming is a highly challengeable task in the face of climate change. In a sustainable farming system, it is expected to keep the continuity in practising of farming throughout a year by introducing autonomy in management of farming system, even under adverse climatic changes or non-availability of feedstock from other sources. In sustainable farming system, the needs of one element are met by the wastes of another component. For example, animal manure builds the soil, replenishing nutrient used by crops that are fed to animals. Livestock farming can be fully autonomous by linking the farmhouse with a polyhouse or greenhouse where variety of feedstock can be developed, even unaffordable climatic conditions (Fig. 6.1). The same way independent from conventional energy, goat manure-based biogas generating system can be developed for fulfilling the energy need of a livestock farmhouse.

The livestock sector is having a clustered of activity of farmers, agro-industry processers and buyers. They must themselves find out ways and means to bring sustainability in farming system. Besides, this marketing has a significant role in sustainability of livestock farming. However, frequent changes in climatic condition, natural disasters, political instability and civil war are major unfavourable factors responsible for food insecurity. This was confirmed by the Food and Agriculture Organization (FAO) of the United Nations, whose main target is to bring global understanding to reduce hunger and promote sustainable agriculture, forestry, fisheries and livestock farming.

It is apprehended that the intensification of global population will mainly take place in developing countries and least developing countries. For example by 2050, India will have 1.73 billion people, Nigeria 401 million, Indonesia 327 million, Pakistan 403 million, Bangladesh 265.8 million, Brazil 232 million, Democratic Republic of Congo 195.3 million and Ethiopia 188.5 million. The increasing trend in rise in urbanisation may demand for food for growing urban population. So, we may face big challenge for increasing agriculture productivity which will take place in villages of developing countries. So, it is high time to highlight the role of livestock sector as key to agriculture development and food security. It is estimated that

B. K. Behera et al., *Move Towards Zero Hunger*,
https://doi.org/10.1007/978-981-32-9800-2_6

Fig. 6.1 Polyhouse for cropping (livestock feedstock)

600 million poor people, mainly consisting of women, manage their livelihood from livestock farming.

6.1 Conceptual Development of Sustainable Livestock Farming

It is obvious that livestock farming has high impact on food security. But we must be aware of the negative effects of the livestock sector. The divesting phenomenon of greenhouse gas emission, overgrazing, water consumption and growing antimicrobial resistance caused due to livestock farming should be mitigated by introducing sustainability in livestock farming practices.

Sustainable livestock farming can bring resource efficiency, strengthens resilience and secures social responsibility of agriculture and food system to ensure nutrition as well as food security to all. Animal agriculture is complementary to sustainable development of a nation [1]. Sustainable key factors are water conservation, soil conservation, soil nutrients management, energy recycling system and biodiversity conservation. Sustainable livestock management is closely associated with energy flow, soil essential element conservation and recycling and continuity of biodiversity.

The various beneficial aspects of livestock farming projects are development of global agricultural economy, rural livelihood upgradation, rural women's empowerment, sustainable rangeland management, conservation of domestic animals and enhancing soil fertility and nutrient cycling. As per the report of FAO [2], the grazing land is more than the cropping land. The grazing land covers about 60% of the total agriculture land and supports 360 million cattle and 600 million sheep and goats. About 10% of the world's beef meat and 30% sheep and goat's meat are produced from grazing animal, worldwide. In addition, sustainable livestock farming not only saves the land for cropping but also is helpful in maintaining nitrogen and phosphorus balance. The goat's excreta are rich in nitrogen and phosphorus. By adapting biogas production techniques from ruminant farm organic wastes, methane gas can be produced to meet the energy demand of farmhouse, and the slurry leftover after biogas production can be used as bio-fertiliser for foliage feedstock production for animals (Fig. 6.2).

In a sustainable livestock farming system, one can monitor the feed conversion efficacy (FCE). It is a biological process to calculate the amount of products in terms of milk, meat and other goat derivatives produced per unit of feed supply to an animal [3]. The efficiency can be increased by application of new technologies in converting feed energy to milk or meat energy. The individual consumes feed as source of energy; the source of energy is converted into different products such as meat, milk and fibre in livestock. The energy gained by an individual is used for body maintenance such as digestion, metabolism, growth and immune function. After fulfilling the body requirement, the energy is channelised for milk or fibre production.

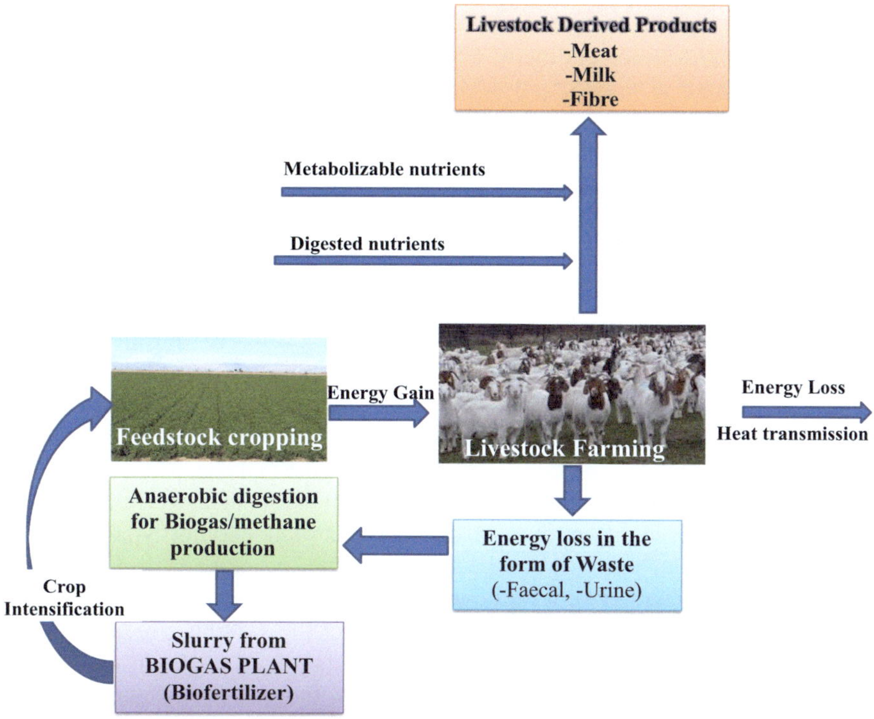

Fig. 6.2 Crop intensified sustainable livestock farming

Maintenance energy requirement is assumed not to change as a function of production. By increasing the caloric value and nutrient value of feed, both the quality and quantity of milk and meat can be increased. The average milk production in the United States was increased from 2074 kg/cow in 1944 to 9193 kg/cow in 2007, and feed input per kg milk was reduced by 77% at 2007. Similarly average daily growth rate of beef cattle in the United States was increased from 0.75 (1977) to 1.08 (2007) kg per head, and feed input per kg grain was reduced by 19% at the 2007 level compared with the 1977 level [4, 5]. Efficiency of ruminant production system is necessary by optimising feed utilisation and monitoring energy losses.

In developing countries, the agriculture productivity is improved by intensification of cropping system with modern agriculture technologies, and cropping intensifies by only 20% from expansion of arable land. By use of "sustainable intensification", the productivity can be more intensified, even under adverse conditions. Sustainable intensification (SI) is a practice for better productivity without the conversion of non-agricultural land. But the concept of SI does not cohere or privilege any particular vision or method of agricultural production. Mostly the term SI is used for various practices like organic farming to minimise adverse environmental impacts like pesticides application, soil erosion and greenhouse gas emission.

By modernisation of agriculture technology and increased awareness for food security, about one-tenth of the world's farmland is under some form of sustainable intensification. This practice can be helpful in improving productivity, conserving biodiversity and ecosystem sustainability. So, it is necessary for government to frame strong policies and peruse the NGOs to bring the beneficial aspects of SI and support the nation in increasing productivity and food security. Livestock farming, especially the farming of ruminant's domestic animals, can convert cheap and commonly available plant resource into high-quality human edible food. It is only the ruminants that can increase quality food supply by exploiting inedible biomass to produce edible protein [6, 7]. It gives an idea about the food use efficiency. Ruminant system is more efficient as compared to non-ruminant system on the basis of input and output edible to human [3]. Animal products are relatively rich in high value energetic proteins such as vitamin B12, vitamin D and zinc [2]. The practice of conventional type of livestock production is in the process of shifting to sustainability intensive practice for more production and food security [8]. About 30–40% of cereals grown globally are used as feed for livestock [9]. Ruminant systems are more efficient when compared with non-ruminant system on the basis of inputs and outputs edible to human [3]. Small ruminant (like goat and sheep) farming system is the best way and means for the better utilisation of marginal lands with prevalence of pastoral system, low level of mechanisation and production of typical products like cheese, high-quality meat and therapeutic milk proteins. Even small ruminants are more efficient feed converters as compared to big ruminants when supplied with coarse forage with low nutrient content. Because of the difference in the milk production level existing between dairy and nondairy breeds of goat, it is necessary to be careful in selecting animal for sustainable goat farming [10]. Nondairy breeds are capable of milk production above the needs of their kids, if they are fed sufficient feeds.

6.2 Types of Sustainable Livestock Farming Systems

Livestock farming system can be categorised on the basis of the sustainable intensification factors like organic livestock farming system, energy autonomous livestock farming system and intergraded livestock farming system.

6.2.1 Organic Livestock Farming

Due to the increase in environmental pollution, biological magnificence of pesticides, hazardous heavy metals and toxic chemicals from industries is in increasing order. Autotrophies are the primary site of pollutants producers. Mostly domestic ruminants like cow, sheep and goat are dependent on green foliage, cereals and dry crops (straws). The residual pesticides, heavy metals and other toxic elements get into the animal system, and human being as the top consumer becomes the ultimate target for such residual toxic chemical. So in order to overcome such problem, the concept of organic farming has started (Fig. 6.2). Organic farming is a method of crop and livestock production that avoids the use of pesticides, fertilisers, antibiotics and growth hormones. By such practice sustainable intensification gets increased and resulted in increasing the productivity both by quality and quantity. Organic farming is a holistic approach in which sustainability of livestock is guaranteed. By this practice one can manage the nutrient value of soil and also improve the soil microflora to maintain nitrogen and phosphorus level of the soil.

Thus, in organic livestock farming, the feedstock is provided with zero contamination of pesticides and other harmful chemicals. So, both the sustainability and the productivity of the farming system depend on internal flow of nutrients in the form of feed and manure, which are mainly for the well-being of livestock.

6.2.2 Energy Autonomy in Sustainable Livestock Farming

In order to bring sustainability in livestock farming system, the rule of energy is of high significance. Most of the rural areas of developing countries are lacking conventional type of electric supply system. This is mainly due to inadequate hydroelectric generation potential of developing countries. The livestock farming system needs continuous energy flow system to bring sustainability in livestock farming in rural areas which are oddly distributed around the peripheral parts of developing countries. On the basis of geographical location, rural areas can generate different sources of nonconventional energy. Renewable energy and livestock farming complement each other in bringing sustainability in the system. Wind, solar and biomass energy can be harvested to provide long-term supply of electricity not only for livestock farming but also for other agricultural activities (Fig. 6.3).

Animal production is basically an input-output system to which law of conservation of energy applies in the same way as it does for any other energetic system. Energy in output (production, losses) requires an equal amount of energy input. It

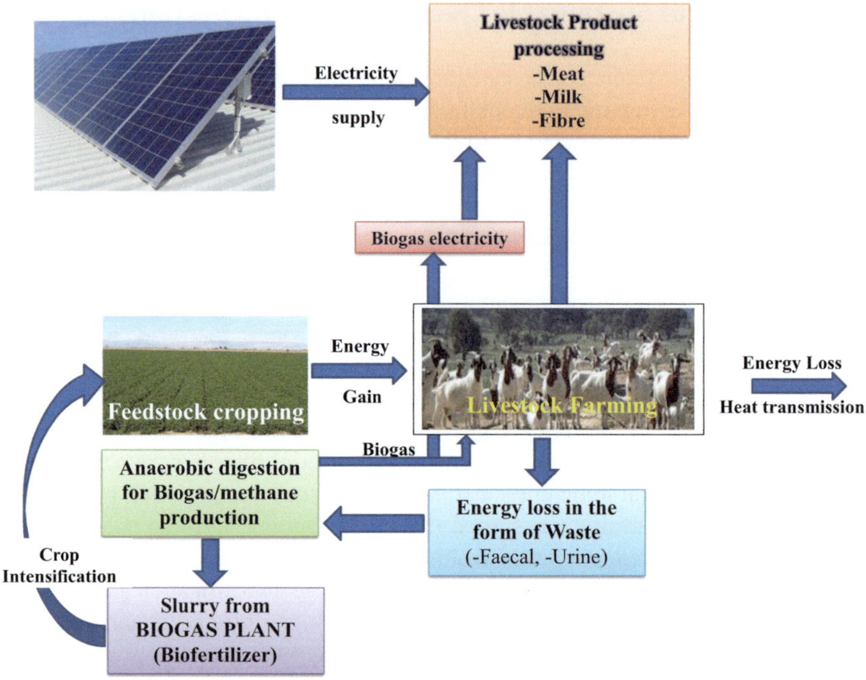

Fig. 6.3 Sustainable livestock farming with energy autonomy

has been suggested that less fossil fuel will be available for animal production by 2050, and different forms of energy should be explored to be used for animal agriculture. It has been estimated that 35 KJ of fossil energy were used to produce 1 kg of feedlot beef [11].

6.2.3 Mixed Livestock Farming

Mixed goat farming is a self-supportive ecosystem which was based on symbiotic relationship with different components. It is the best way of energy cycling and increases overall production and enhanced nutrient supply. The system provides energy security, soil and water conservation, better utilisation of space and increasing cropping intensity and manages pest incidence. This makes the farming system resilient and highly productive. In integrated farming soil fertility remains intact due to sustainable pattern of natural resource development process.

Mixed farming system is the best way of sustainable resource management. In this process crop, pond, garden and animal resources management system is integrated for better production and to bring sustainability. The animals such as cow, goat, rabbits, pig and chicken can be introduced to provide waste products as source of nutrients and other functional inputs. The system provides effective linkages between various components, thereby reducing cost of production, effort and risk and enhancing diversity and nutrition.

Model: Agroforestry + goat + horticulture
Rice + cow + chicken

Farmers can analyse seasonal variation; available resources such as farmland, homestead, waterbed and livestock; and food processing and marketing.

Model:

- Mixed cropping – Pea/Bengal gram/Rice
- Vegetable + grain
- Livestock +fish pond + horticulture
- Vermicompost
- Livestock (goat, chicken) + crop (rice) + fish (pond)
- Livestock (small ruminant) + crop + horticulture
- Livestock (small ruminant) + silvipasture
- Livestock + crop

The use of cow dung, goat fecal material helps in growth of phytoplankton and zooplanktons in fish pond. The livestock system by-products manure, urine and spilled feed.

- Livestock + crop + backyard goat or poultry

This increases nutritional security and better income.

- Sheep/goat + crop + horticulture based farming system
- Small ruminant + silvi-pastoral based farming system
- Livestock + crop

Livestock-fish integration maximises food production and economic return per unit area of land. Its main benefit comes from the conversion of livestock wastes (faeces, urine, stall/pen washing, spilled animal feed) into protein. More important is the indirect fertilisation effects of livestock wastes entering the pond, thus enhancing microbial and plankton production which contributed to fish nutrition. Integrating goat farming with cropping systems like agroforestry models such as agri-silviculture, hortipasture, boundary plantations, wind belts and silvipasture models with trees and grasses can be established in wastelands to supplement farm income. The plant commonly used as goat fodder in specific region should be planted for continuous fodder supply economises the cost of feed and fodder.

6.3 Production Efficiency and Emissions

Livestock farming has a vast environmental footprint. In spite of its beneficial effect, livestock farming, if not properly managed, has an alarming impact on nature's conservation. The main problem with livestock farming is emission of methane to atmosphere from the solid organic waste by anaerobic bacteria decomposition. About, 18% of greenhouse gas contributes to atmosphere is from unusual management of livestock farming. Besides this, it contributes to land water degradation, biodiversity loss, acid rain, coral reef degradation and deforestation. Ruminants have lower efficiency of N utilisation and higher excretion of N in faeces and urine per unit N intake compared with non-ruminants [12].

6.3.1 Greenhouse Gas Emissions Sources from the Livestock

The worst important greenhouse gases from livestock sources are methane, nitrogen and nitrous oxide. The demand for livestock products is increasing in developing countries also. Therefore the livestock sector has to follow strategy to reduce environmental burden and simultaneously fulfil the food demand for the growing world population.

The environment emission from different producers is composed of carbon dioxide equivalent. The carbon dioxide equivalent is the standard unit used to account for global warming potential [13]. The greenhouse gas emission in livestock production comes from enteric fermentation, manure storage and feed production.

Enteric Fermentation

Enteric fermentation occurs due to breakdown of biomass in rumen, due to action of bacteria, protozoa and fungi. Ruminants convert cellulose and semi-cellulose into energy need of individuals. Plant biomass in the rumen is converted into volatile fatty acids, which pass the rumen wall and go to the liver through the circulatory system. Enteric fermentation produces about 2.8 gigatonnes and produces about 39% of total emission.

Manure Storage

Manure acts as an emission source for both methane and nitrous oxide. The emission depends on type of management and composition of the manure. Manure storage produced about 0.71 gigatonnes and accounts for about 10% of total emission. Methane and nitrous oxide production in manure depends on organic matter and nitrogen content of excreta. Under anaerobic conditions, the organic matter is partially decomposed by bacteria producing methane and carbon dioxide. When manure is stored as a solid (dung) or deposited in pastures, nitrous oxide is generated through both the nitrification and denitrification process of the nitrogen contained in manure.

Feed Production

Feed production includes all the greenhouse gas emission arising from the following activities such as:

- Land use change
- Manufacturing and use of fertilisers and pesticides

- Manure excreted and applied to fields
- Agricultural operations
- Feed processing
- Feed transport

Feed production accounts for about 45% of the total emission of whole livestock sector and is about 3.2 gigatonnes of carbon dioxide equivalent.

6.3.2 Methane Efficiency

Methane is formed by methanogenic archaea in rumen. Methane is mainly formed in rumen (90%), and about 10% of methane is formed in hindgut. Methane production depends on various factors such as composition of diet and level of feed intake in the ruminant. Ruminant production system with higher intake of fibrous feed may be associated with high levels of GHG emission per unit products.

Ruminant consumes human inedible feed (fibre-rich forages and by-products), convert it into human edible food, and produces CH_4 in this process. The consumption of fibrous feeds is associated with high levels of GHG emissions per unit product and indirectly discouraging us not to consume beef. The enteric CH_4 emission can be reduced by different dietary options [14–16]. The addition of dietary fat or oil reduces CH_4 production by about 5% with 1% addition of lipid to the diet [17].

Dietary lipid supplementation decreases CH_4 production by reducing the activity of methanogens and protozoal numbers and reducing fermentation of substrate in the rumen. Monensin, a rumen modifier, reduces CH_4 production, but the use of monensin in livestock industry is not allowed. Yeast culture is used to maintain the milk yield and production efficiency and thus may decrease CH_4 production. Nitrate is also highly effective in reducing CH_4 in ruminant, but the risk of nitrate poisoning is important [18]. Besides, this number of dietary additives had shown promising effect on CH_4 production in vitro and did not show effectiveness in vivo.

As the production system becomes efficient, the waste excretion decreases per unit product. The N and CH_4 efficiency are more observed at low level of production. The ability of ruminants to convert human inedible products into human edible products is an important concern with respect to global food security. The ruminant animals are mainly dependent on natural resource for their food. Animals like goat, sheep and cow are the cheap food human consumption, and helpful in achieving food security. But intensive production of animal system has become source of N pollution and GHG emission.

6.3.3 Mitigation

Sustainable management of livestock system including agricultural sector is necessary to reduce emission potential. Forage quality and digestibility affect enteric methane production. The practice to increase forage quality reduces enteric methane production of about 5% per unit of fat-corrected milk [19]. The processing of forages such as chopping and grinding improves digestibility and mitigates enteric methane production [20]. The concentrate supplementation also increases feed digestibility in rumen and reduces 15% of methane emission per unit of fat-protein-corrected milk [19]. The ratio of forage to concentrate has to be provided strategically to increase digestibility and reduce methane. The higher proportion of carbohydrate could affect rumen digestibility and increase the risk of metabolic diseases. Similarly lipid diet supplementation between 5% and 8% of the dry matter intake is an effective mitigation strategy with a potential enteric methane reduction of about 15% per unit of fat-protein-corrected milk [17, 19].

Frequent removal of manure from shed is an effective practice to reduce methane emission. Cattle and pig farmers should clean the shed and store the manure in an outside storage facility. The poultry farm should be careful in waste management and timely disposal process. Accumulation of such waste may cause emission of methane gas and other nitrous gas. Potential grazing system is practised as a good approach to increase forage production and reduce nitrous oxide emissions. By rotating the animals in the available area, farmers can manage stocking densities and grazing duration and thereby manage nitrogen excreta distribution and vegetation regrowth. The urine is also distributed uniformly in paddock area and reduces effective nitrogen application rate, thereby reducing nitrous oxide emissions [21]. The animals should be kept away from paddock during wet weather to reduce soil compaction and sward damage. Moreover, avoiding excreta deposition during the period will reduce nitrous oxide emissions and nitrogen leaching [22]. Breeding for more productive animals will increase the feed conversion efficiency and lead to reduction of the nutrient requirements. A more efficient animal will retain more dietary nitrogen protein, and there will be less nitrogen in faeces and urine [23].

Management practices significantly affect the greenhouse gas production. Poor livestock health and welfare are associated with behavioural and metabolic changes and thereby affecting greenhouse gas emission up to 24% per unit milk produced and up to 113% per unit beef carcass [24].

6.4 Pasture Land Sustainability

Pasture land is the farmland cover and used for animal production and comprising about 22–26% of earth's ice-free land surface. Land used for animal production is supporting dietary needs worldwide and screening livelihoods for nearly 1.3 billion people [25]. Pasture or rangeland is the most extensive form of used land and comprising about 22–26% of the earth's ice-free land surface [14, 26]. Pasture or rangeland is commonly used for grazing of animals globally. Land used permanently for herbaceous forage crop either by cultivated or natural grazing [27]. Moreover, there are some browse lands that contain shrubs, trees and bushes and are used for animal production. The sustainable livelihood systems are hunting and gathering, animal rearing, cultivation and fishing. Pasture land refers to graminoids, herbs and forages that may undergo animal grazing. Browse land refers to trees, shrubs and succulents that may undergo browsing. Mixed agriculture land refers to land that is mainly used for animal production and cultivation. Crop land includes foraging through consumption of residues. In animal production system, the grazing animals use mostly graminoids (grasses, sedges), herbs and forbs; the browsing animals use trees, shrubs and succulents on browse lands; and foraging animals use lichens, mushrooms, nuts and fruits on pasture and browse land.

The land use pattern has changed dramatically as the percentage of total land area used for crop land and pasture has expanded at the expense of natural habitats [25]. The world population is expected to grow from 6.9 billion in 2010 to 9.3 billion in 2050 [28]. Therefore resources such as arable land, water, minerals and fuel should be used optimally to maintain sustainable production.

6.5 Developing Family Farming

There are about 570 million farms in the world; China and India represent 35% and 24% of 570 million farms, respectively. There are about 9% farms in sub-Saharan Africa and 7% in Europe and Central Asia. There are only 3% of the world farm that are located in the Middle East and North Africa.

Family farming (which includes all family-based agricultural activities) is a means of organising agricultural, forestry, fisheries, pastoral and aquaculture production which is managed and operated by a family and predominantly reliant on family labour, including both women's and man's. The family and the farm are linked, co-evolve and combine economic, environmental, social and cultural functions [29]. Family farming (FF) includes agricultural, forestry, fisheries, livestock and aquaculture production which is managed and operated by a family and predominantly reliant on family labour. Family farming contributes significantly to the world's food production both in developed and developing countries [30] and provides livelihood to 80% of the world's poor living in rural areas. Nagayets [31] reported that the average farm sizes decreased globally from 1960 through 1990. FAO [29] also reported that the average farm size decreased from 1950 to 2000 at global level. Family farming plays an important role in alleviating hunger and poverty, providing food security and nutrition, improving livelihoods, sustainably managing natural resources, protecting the environment and fostering sustainable development.

Family farming is a key to enhanced food security, sustainable growth and the fight against rural poverty and environmental degradation. Family farming fights with hunger by strengthening rural farm economy. Family farming reduces hunger and poverty. It also maintains local traditions, heritage and food systems, community ecosystems and rural landscapes. Family farming produces about 80% of the world's food and simultaneously serves as custodians of about 70–80% of farmland. Family farmers address the root cause of rural poverty, food insecurity, migration and environmental degradation. Family farmers mainly work in multicropping farming systems. Multicropping farming system enhances resilience to crop failures and price shocks and contributes towards improved food and nutritional security.

Multicropping systems are more adaptable to low-carbon agriculture, as they are less fossil fuel dependent for production and transport of fresh food to local markets. Family farmers have the potential to promote environmental sustainability of agriculture.

Rural families, particularly women, contribute to family farming with their knowledge and labour and in managing agricultural biodiversity. Women work both at home and field and work as major catalyst for the diversification of income source. Despite their essential contribution, women have less access than men to productive resources and opportunities – land, livestock, labour, education, extension and financial services and technology – which limits their capacity to generate marketable surpluses and add value to their production. Family farmers constitute about 43% of the agricultural labour force in developing countries, ranging from 20% in Latin America to 50% in Eastern Asia and sub-Saharan Africa.

Most of the rural poor are small-scale family food producers who depend on agriculture and aquaculture for their food and income but face many difficulties accessing productive resources, opportunities and markets. Family farmers include pastoralist, fisherfolk and mountain farmer. Fisherfolk supplies 60% of the fish for direct human consumption [30]. Approximately 200–500 million pastoralists herd their animals on rangelands that cover one-third of the earth's land surface and derive sustenance from extensive nomadic, semi-nomadic and transhumant livestock rearing. They produce useful animal products in the world's harshest environments and contribute significantly to food security in these regions [32]. Mountain farming is largely family farming. Around 40% of mountain populations in developing and transition countries – about 300 million people – are food insecure, with half of them suffering from chronic hunger [33]. Family farmers include forest communities. Around 40% of the extreme rural poor live in forest and savannah areas [34]. The unique combination of forest and farm resources has created complex natural resource management systems all over the world [35].

6.6 Livestock Linked to Bioaccumulation of Antibiotics

The emergence and spread of agricultural antibiotics are matter of concern, and "farming practices are largely to blame for the rise of antibiotic-resistant strains" [36]. A human is infected by a resistant pathogen of agricultural origin through contact with livestock or through ingestion of bacteria from contaminated meat or water, without ongoing transmission of the pathogen between humans. Resistance genes arising in the agricultural setting are introduced into human pathogens by horizontal gene transfer. The resulting resistant lineages are then selected by antibiotic use in humans.

Animal husbandry is a major food-producing industry worldwide. Poor gut health is associated with infection, inflammation and intestinal atrophy leading to stunted growth, morbidity and mortality in livestock. Antibiotics are mainly used in livestock industry to prevent the diseases as well as growth promoter. Global consumption of antibiotics from food animal production was 63,151 tons in 2010, and the predicted estimate will rise to 67% by 2030 [37]. The practice has been especially linked to antimicrobial-resistant pathogens in both livestock and humans. Due to widespread use of antibiotics for treatment, prophylaxis and growth promotion, livestock has become a reservoir of antimicrobial-resistant bacterial strains and genes. The antimicrobial-resistant strains are transmitted directly to farmers and veterinarians and to consumers through food chain [38, 39]. The antibiotic residues in excrement may cause high level of antibiotic accumulation in the environment.

The livestock is showing increasing antimicrobial resistance in similar manner to human clinical isolates [40]. Antibiotics are not degraded in the animal body and thereby enter into environment through excreted urine and faeces [41]. The livestock farms have been reported to discharge a large amount of antibiotics into nearby surface water and soil [42, 43]. Residual antibiotics usually have negative impacts on organisms, food security and water quality, which in turn accumulate in the human body through the food chain and drinking water [44–46]. The problem is evident in both developed and developing countries. The maximum concentration of oxytetracycline was 340 ng/L in the United States [47]; and similarly the maximum concentrations of sulfadiazine and oxytetracycline were 4130 ng/L and 4490 ng/L, respectively, in the United Kingdom [48].

Antimicrobials are delivered to animals for a variety of reasons including disease treatment, prevention and control and growth promotion/feed efficiency. Antimicrobial growth promotants (AGPs) were first advocated in the mid-1950s, when it was discovered that small, subtherapeutic quantities of antibiotics such as procaine penicillin and tetracycline (1/10–1/100 the amount of a therapeutic dose) delivered to animals in feed could enhance the feed-to-weight ratio for poultry, swine and beef cattle [49]. For many years, the positive effects of this practice were championed, while the negative consequences went undetected. But microbiologists and infectious disease experts facing antibiotic resistance questioned the possible harm from this use [50–53]. They found that in farms using AGPs the intestinal floras of the farm workers and farm animals had more resistant bacteria than in those people and animals on farms not using AGPs. A prospective in vivo/in situ study in 1975 was performed to evaluate the effect of introducing low-dose in-feed oxytetracycline as an AGP on the intestinal floras of chickens and farm dwellers [54]. The results showed not only colonisation of the chickens with tetracycline-resistant and other drug-resistant *Escherichia coli* strains but also acquisition of resistance of *E. coli* in the intestinal flora of the farm family. Other studies over the ensuing three decades further elucidated the quantitative and qualitative relationships between the practice of in-feed antimicrobials for animals and the mounting problem of hard-to-treat, drug-resistant bacterial infections in humans [55, 56]. Antimicrobial resistance poses a worldwide threat to public health, which may be partially associated with using AGP in livestock production. The European Union banned the use of AGP in animal food production in 2006. The US FDA placed restrictions on antibiotic use in animal production in December 2016.

6.6.1 Nutrient Enrichment of Gut

It has been proven that the gut microbiome has enormous functional potential for the host such as human being. The diet plays a fundamental role in shaping the composition of gut microbiota and therefore determines the interrelationship between the gut microbiome and host (human being) [57]. It has been shown that microbiome composition and functions are responsible for human metabolism and regulate balance between health and disease. The microbiome diversity is associated with triggering of various biochemical pathways, thereby increasing proper intestinal permeability and immunomodulation. The microbiota composition is a key factor affecting responsiveness for different nutrition status.

The manipulation of gastrointestinal microbiota in maintaining animal gut health can be maintained by probiotics and prebiotics. Probiotics can manage the prevention of gastrointestinal infection and disease in an effective manner. The probiotics modulate the gastrointestinal environment and modulate the immune system of the host. It has been acknowledged that the beneficial effects of probiotics are shared among a vast number of genera and species belonging not only to the human or animal gastrointestinal tract. Probiotics have been used since many years without causing any problem but have to be chosen carefully.

Probiotics

The term "probiotic" comes from two Greek words "pro" and "bios" meaning "for life". Mechnikov in 1907 probably suggested the first concept of probiotics and noted that bacteria may have a beneficial effect on the natural gut microbiota. FAO and WHO working group formulated the definition of *probiotics* in 2002 as "live strains of strictly selected microorganisms which, when administered in adequate amounts, confer a health benefit on the host". The definition was maintained by the International Scientific Association for Probiotics and Prebiotics (ISAPP) in 2013. Proper selection of bacterial strains and application of a correct dose are recommended for efficacy of probiotics. Probiotics have a number of effects in the body, and different probiotics act in different ways. The ideal probiotic should be non-pathogenic (GRAS), acid and bile tolerant, genetically stable, lactic acid producer, anti-genotoxic, effectively adhesive to gut lining, generating short time, robust and surviving processing conditions. The other beneficial aspects of probiotic include:

- Facilitate to maintain a desirable community of microorganisms.
- Soothe and stabilise the digestive tract's barriers against undesirable microbes in GIT or produce elements that inhibit their growth.
- Help the community of microorganisms in the digestive tract to return to normal after being disturbed (e.g. by an antibiotic or a disease).
- Eliminate undesirable microorganisms, viz. bad bacteria, yeast and fungi.

- Stimulate the immune response by stimulating secretion of IgA and regulatory T cells.
- Produce vitamin B12, butyrate and vitamin K2.

The popularly used probiotic microorganisms are *Lactobacillus rhamnosus*; *Lactobacillus reuteri*; *Bifidobacteria*; certain strains of *Lactobacillus casei*, *Lactobacillus acidophilus* group and *Bacillus coagulans*; *Escherichia coli* strain Nissle 1917; certain enterococci especially *Enterococcus faecium* SF68; and the yeast *Saccharomyces boulardii*. Bacterial spore formers, mostly of the genus *Bacillus*, dominate the scene. These probiotics are added to foods, particularly fermented milk products, either singly or in combinations. New genera and strains of probiotics are continuously emerging with more advanced and focused research efforts.

Prebiotics

Prebiotics are used as natural feed additives. The prebiotic concept was first initiated in 1995. The definition says that the group of prebiotics may involve other substances besides carbohydrates (such as polyphenols and polyunsaturated fatty acids transformed into corresponding conjugated fatty acids) and may act not only in the alimentary tract. Another important aspect is that they are no longer limited to human food but may be also considered in other categories, such as animal nutrition. Prebiotics are mostly fibres that are non-digestible food ingredients and beneficially affect the host's health by selectively stimulating the growth and/or activity of some genera of microorganisms in the colon, generally lactobacilli and bifidobacteria [58].

An ideal prebiotic (1) should be resistant to the actions of acids in the stomach, bile salts and other hydrolysing enzymes in the intestine, (2) should not be absorbed in the upper gastrointestinal tract and (3) should be easily fermented by the beneficial intestinal microflora [59]. Some of the sources of prebiotics include breast milk, soybeans, inulin sources (like Jerusalem artichoke, chicory roots, etc.), raw oats, unrefined wheat, unrefined barley, yacon and non-digestible carbohydrates, in particular non-digestible oligosaccharides. However, among prebiotics only bifidogenic, non-digestible oligosaccharides (particularly inulin, its hydrolysis product oligofructose and (trans) galacto-oligosaccharides (GOS)) fulfil all the criteria for prebiotic classification [60].

Prebiotics like inulin and pectin exhibit several health benefits like reducing the prevalence and duration of diarrhoea, relief from inflammation and other symptoms associated with intestinal bowel disorder and protective effects to prevent colon cancer [61]. They are also implicated in enhancing the bioavailability and uptake of minerals, lowering some risk factors of cardiovascular disease and promoting satiety and weight loss, thus preventing obesity [60].

Synbiotics

In 1995, Gibson and Roberfroid introduced the term "synbiotic" by specifying in this way: "a mixture of probiotics and prebiotics that beneficially affects the host by improving the survival and implantation of live microbial dietary supplements in the

GI tract, by selectively stimulating the growth and/or activating the metabolism of one or a limited number of health-promoting bacteria, and thus improving host welfare" [62]. As the word "synbiotic" implies synergy, the term should be reserved for those products in which a prebiotic component selectively favours a probiotic microorganism [63]. The principal purpose of that type of combination is improvement of survival of probiotic microorganisms in the gastrointestinal tract. Synbiotics have both probiotic and prebiotic properties and were created in order to overcome some possible difficulties in survival of probiotics in the gastrointestinal tract [64]. Probiotics beneficially influence the intestinal equilibrium and constitute a protective barrier for the alimentary tract. Prebiotics, on the other hand, supply energy and nutrients for probiotic bacteria [65, 66]. Therefore, an appropriate combination of both components in a single product should ensure a superior effect, compared to the activity of the probiotic or prebiotic alone [67, 68]. Synbiotics were developed to overcome possible survival difficulties for probiotics. It appears that the rationale to use synbiotics is based on observations showing the improvement of survival of the probiotic bacteria during the passage through the upper intestinal tract. A more efficient implantation in the colon and a stimulating effect of the growth of probiotics and ubiquitous bacteria contribute to maintain the intestinal homeostasis and a healthy body [61].

Health Benefits of Probiotics, Pre-probiotics and Synbiotics

The most important and documented beneficial effects of probiotics include the prevention of diarrhoea and constipation, changes in bile salt conjugation and enhancement of antibacterial and anti-inflammatory activity. Furthermore, they also contribute to the synthesis of nutrients and improve their bioavailability. Probiotics have also demonstrated their inherent effects in alleviating symptoms of allergy, cancer, AIDS and respiratory and urinary tract infections. The number of mechanisms are involved in probiotics beneficial effects such as: production of inhibitory substances like H_2O_2, bacteriocins, organic acids; blocking of adhesion sites for pathogenic bacteria; blocking of toxin receptors and modulation of immune responses.

References

1. Raman S (2006) Agricultural sustainability: principles, processes, and prospects. Food Products Press, an imprint of The Haworth Press, Binghamton
2. FAO (2011) World livestock 2011 – livestock in food security. FAO, Rome
3. Reynolds MP, Hellin J, Govaerts B et al (2012) Global crop improvement networks to bridge technology gaps. J Exp Bot 63(1):1–12
4. Capper JL, Cady RA, Bauman DE (2009) The environmental impact of dairy production: 1944 compared with 2007. J Anim Sci 87:2160–2167
5. Capper JL (2011) The environmental impact of beef production in the United States: 1977 compared with 2007. J Anim Sci 89:4249–4261
6. Gill M, Smith P, Wilkinson JM (2010) Mitigating climate change: the role of domestic livestock. Animal 4:323–333
7. Wilkinson JM (2011) Re-defining efficiency of feed use by livestock. Animal 5:1014–1022

8. Gerber PJ, Vellinga TV, Steinfeld H (2010) Issues and options in addressing the environmental consequences of livestock sector's growth. Meat Sci 84:244–247
9. Godfray HCJ, Beddington JR, Crute IR et al (2010) Food security: the challenge of feeding 9 billion people. Science 327(5967):812–818
10. Sands M, McDowell RE (1978) The potential of the goat for milk production in the tropics. Cornell University, Department of Animal Science, Ithaca, International Agricultural Mimeo 60
11. Hillel D, Rosenzweig C (2008) Biodiversity and food production. In: Chivian E, Bernstein A (eds) Sustaining life: how human health depends on biodiversity. Oxford University Press, New York, pp 325–381
12. Kohn RA, Dinneen MM, Russek-Cohen E (2005) Using blood urea nitrogen to predict nitrogen excretion and efficiency of nitrogen utilization in cattle, sheep, goats, horses, pigs, and rats. J Anim Sci 83:879–889
13. IPCC (2013) Summary for policymakers. In: Stocker TF, Qin D, Plattner GK et al (eds) Climate change 2013: the physical science basis, Contribution of Working Group I to the Fifth Assessment Report of the Intergovernmental Panel on Climate Change. Cambridge University Press, Cambridge/New York, p 1535
14. Ellis JL, Bannink A, France J et al (2010) Evaluation of enteric methane prediction equations for dairy cows used in whole farm models. Glob Chang Biol 16:3246–3256
15. Beauchemin KA, McAllister TA, McGinn SM (2009) Dietary mitigation of enteric methane from cattle. CAB Rev: Perspectives Agric Vet Sci Nutr Natur Resour 4:035
16. Martin C, Morgavi DP, Moreau D (2010) Methane mitigation in ruminants: from microbe to the farm scale. Animal 4:351–365
17. Grainger C, Beauchemin KA (2011) Can enteric methane emissions from ruminants be lowered without lowering their production? Anim Feed Sci Technol 166–167:308–320
18. Nolan JV, Hegarty RS, Hegarty J et al (2010) Effects of dietary nitrate on rumen fermentation, methane production and water kinetics in sheep. Anim Prod Sci 50:801–806
19. Knapp JR, Laur GL, Vadas PA et al (2014) Invited review: enteric methane in dairy cattle production: quantifying the opportunities and impact of reducing emissions. J Dairy Sci 97:3231–3261
20. Hristov AN, Oh J, Firkins JL et al (2013) Mitigation of methane and nitrous oxide emissions from animal operations: I. a review of enteric methane mitigation options. J Anim Sci 11(91):5045–5069
21. Eckard RJ, Grainger C, de Klein CAM (2010) Options for the abatement of methane and nitrous oxide from ruminant production: a review. Livest Sci 130:47–56
22. Luo Y, Wei Z, Sun Q et al (2011) Effects of zeolite addition on ammonia volatilization in chicken manure composting. Nongye Gongcheng Xuebao/Trans Chin Soc Agric Eng 27:243–247
23. Gerber PJ, Steinfeld H, Henderson B et al (2013) Tackling climate change through livestock – a global assessment of emissions and mitigation opportunities. Food and Agriculture Organization of the United Nations (FAO), Rome
24. Williams A, Chatterton J, Hateley G et al (2015) A systems-life cycle assessment approach to modelling the impact of improvements in cattle health on greenhouse gas emissions. Adv Anim Biosci 6:29–31
25. Steinfeld H, Gerber P, Wassenaar T et al (2006) Livestock's long shadow: environmental issues and opinions. FAO, Rome
26. Goldewijk KK, Beusen A, Janssen P (2010) Long term dynamic modeling of global population and built-up area in a spatially explicit way, HYDE 3.1. The Holocene 20(4):565–573
27. FAO (2014b) FAOSTAT. Online statistical database (retrieved November 2014) (available at http://faostat.fao.org)
28. United Nations Department of Economic and Social Affairs/Population Division World Population Prospects: The 2010
29. FAO (2013c) Agribusiness public–private partnerships: a country report of Thailand. Rome

30. FAO, SOFA (2014) The State of Food and Agriculture 2014. Innovation in Family Farming. Rome
31. Nagayets O (2005) Small farms: current status and key trends. Prepared for the Future of Small Farms Research Workshop held in Wye, UK. June 26–29. Washington, D.C., IFPRI
32. FAO (2016) The state of food and agriculture 2016. Climate change, agriculture and food security. Rome
33. FAO (2013) 2000 world census of agriculture: analysis and international comparison of the results (1996–2005). FAO Statistical Development Series No. 13. Rome
34. FAO, SOFO (2018) State of the World's Forests 2018. Forest pathways to Sustainable Development. Rome
35. FAO, SOFO (2012) State of the world's forests 2012. Rome
36. Kennedy D (2013) Time to deal with antibiotics. Science 342:777
37. Van Boeckel TP, Brower C, Gilbert M et al (2015) Global trends in antimicrobial use in food animals. Proc Natl Acad Sci USA 112:5649–5654
38. Vishnuraj MR, Kandeepan G, Rao KH et al (2016) Occurrence, public health hazards and detection methods of antibiotic residues in foods of animal origin: a comprehensive review. Cogent Food Agric 2:1
39. Marshall BM, Levy SB (2011) Food animals and antimicrobials: impacts on human health. Clin Microbiol Rev 24:718–733
40. O'Neill J (2015) Rapid diagnostics stopping the unnecessary use of antibiotics. http://www.sfam.org.uk/download.cfm?docid=1191C802-F9CA-4406- 8DA9ED9E3F7B2A7F. 2015.10. Accessed April 6, 2018
41. Lim SJ, Seo CK, Kim TH et al (2013) Occurrence and ecological hazard assessment of selected veterinary medicines in livestock wastewater treatment plants. J Environ Sci Health B 48:658–670
42. Watanabe N, Bergamaschi BA, Loftin KA et al (2010) Use and environmental occurrence of antibiotics in freestall dairy farms with manured forage fields. Environ Sci Technol 44:6591–6600
43. Zhu Y, Johnson TA, Su J et al (2013) Diverse and abundant antibiotic resistance genes in Chinese swine farms. Proc Natl Acad Sci USA 110:3435–3440
44. Sarmah AK, Meyer MT, Boxall AB (2006) A global perspective on the use, sales, exposure pathways, occurrence, fate and effects of veterinary antibiotics (vas) in the environment. Chemosphere 65:725–759
45. Underwood JC, Harvey RW, Metge DW et al (2011) Effects of the antimicrobial sulfamethoxazole on groundwater bacterial enrichment. Environ Sci Technol 45:3096–3101
46. Jia S, Zhang XX, Miao Y et al (2017) Fate of antibiotic resistance genes and their associations with bacterial community in livestock breeding wastewater and its receiving river water. Water Res 124:259–268
47. Kolpin DW, Furlong ET, Meyer MT et al (2002) Pharmaceuticals, hormones, and other organic wastewater contaminants in U.S. Streams, 1999–2000: a national reconnaissance. Environ Sci Technol 36:1202–1211
48. Environmental Agency (2005) Targeted monitoring study for veterinary medicines in the UK environment. Environment Agency, Rio House, Waterside Drive, Aztec West, Almondsbury, Bristol, BS32 4UD. https://www.gov. uk/government/uploads/system/uploads/attachment_data/file/290533/ scho0806blhh-e-e.pdf. Accessed March 13, 2014
49. Stokestad ELR, Jukes TH (1950) Further observations on the "animal protein factor". Proc Soc Exp Biol Med 73:523–528
50. Gorbach SL (2001) Antimicrobial use in animal feed—time to stop. N Engl J Med 345:1202–1203
51. Howells CH, Joynson DH (1975) Possible role of animal feeding stuffs in spread of antibiotic-resistant intestinal coliforms. Lancet:156–157
52. Levy SB (2002) The antibiotic paradox: how the misuse of antibiotics destroys their curative powers, 2nd edn. Perseus Publishing, Cambridge, MA

53. Smith HW, Crabb WE (1957) The effect of the continuous administration of diets containing low levels of tetracyclines on the incidence of drug-resistant Bacterium coli in the faeces of pigs and chickens: the sensitivity of the Bact. coli to other chemotherapeutic agents. Vet Rec 69:24–30
54. Levy SB, FitzGerald GB, Macone AB (1976) Spread of antibiotic resistant plasmids from chicken to chicken and from chicken to man. Nature 260:40–42
55. Hershberger E, Oprea SF, Donabedian SM et al (2005) Epidemiology of antimicrobial resistance in enterococci of animal origin. J Antimicrob Chemother 55:127–130
56. Witte W (2000) Selective pressure by antibiotic use in livestock. Int J Antimicrob Agents 16(Suppl. 1):S19–S24
57. Shanahan F, van Sinderen D, O'Toole PW et al (2017) Feeding the microbiota: transducer of nutrient signals for the host. Gut 66:1709–1717
58. De Vrese M, Schrezenmeir J (2008) Probiotics, prebiotics and synbiotics. In: Stahl U, UEB D, Nevoigt E (eds) Food biotechnology, advances in biochemical engineering/biotechnology. Springer, Berlin, pp 1–66
59. Kuo SM (2013) The interplay between fiber and the intestinal microbiome in the inflammatory response. Adv Nutr 4(1):16–28
60. Pokusaeva K, Fitzgerald GF, van Sinderen D (2011) Carbohydrate metabolism in *Bifidobacteria*. Genes Nutr 6(3):285–306
61. Peña AS (2007) Intestinal flora, probiotics, prebiotics, synbiotics and novel foods. Revista Espanola de Enfermedades Digestivas 99(11):653
62. Gibson RG, Roberfroid MB (1995) Dietary modulation of the human colonic microbiota: introducing the concept of prebiotics. J Appl Bacteriol 125(6):1401–1412
63. Cencic A, Chingwaru W (2010) The role of functional foods, nutraceuticals, and food supplements in intestinal health. Nutrients 2(6):611–625
64. Rioux KP, Madsen KL, Fedorak RN (2005) The role of enteric microflora in inflammatory bowel disease: human and animal studies with probiotics and prebiotics. Gastroenterol Clin N Am 34:465–482
65. Blay GL, Michel C, Blottiere HM et al (1999) Prolonged intake of fructo-oligosaccharides induces a short-term elevation of lactic acid-producing bacteria and a persistent increase in cecal butyrate in rats. J Nutr 129(12):2231–2235
66. Gibson GR (2003) Prebiotics. Best Pract Res Clin Gastroenterol 18:287–298
67. Bengmark S (2005) Bioecological control of the gastrointestinal tract: the role of flora and supplemented probiotics and synbiotics. Gastroenterol Clin N Am 34:413–436
68. Panesar PS, Kaur G, Panesar R et al (2009) Synbiotics: potential dietary supplements in functional foods. Food Science Central

7 Micro Livestock Farming

7.1 Micro Livestock Farming

Micro livestock are the source of income and nutrition for farmers in all parts of the world. Micro livestock primarily ensure nutritional security for small farmers as they manage to rear them in their home or backyard with the available foodstuffs. The landless farmers can afford to raise such micro animals in their backyard under ambient climatic condition. Micro livestock are promising for subsistence production and contribute significantly to family nutrition. Small animals are easy to manage with minimum financial risk and faster return on investment. The space requirement for handling and feeding micro livestock is less. The small animal meat can be consumed by families in one meal or for a whole day to reduce the risk of spoilage where there is no facility for refrigeration.

Small animals require higher input of feed because they grow faster than large animals. They are most efficient converters of food energy. They increase the chances of successful breeding and provide a steady source of income. Small animals tend to fit well into existing farming systems and can thrive well under low nutrient availability of that area. Small animals attain early sexual maturity with short gestation period and produce a number of offspring and have high juvenile growth. Moreover, in remote areas and in areas of extreme climate, they are the major source of basic subsistence. Micro breeds may be the only practical livestock in the situation of less land, less forage, higher stress and diseased environment. The individual output may be low, but it can be efficient by comparing the environmental situation and other inputs. Small animals, particularly non-ruminants, require higher protein feed and lower fibre content per unit of body weight than large animals. Low investment, low and cheap labour requirement, fast return on capital, flexibility and efficient resource utilization are major advantages of rearing micro livestock. Micro livestock has not been given due attention and their potential has not been envisaged to bring livestock revolution. There is a considerable scope of developing micro livestock to improve the income level of resource-poor farmers.

B. K. Behera et al., *Move Towards Zero Hunger*,
https://doi.org/10.1007/978-981-32-9800-2_7

Micro livestock are reared in high density and cause for spreading infectious disease and parasites. Again micro livestock are reservoirs of spreading communicable disease to humans. Micro livestock production should be integrated with different rural develop projects in the areas where no other mechanism is available for improving their living standard. Poor farmers who are raising small animals by nourishing them with the locally available feed and fodder resources produce quality protein to satisfy their basic needs. It is not the question of how small or large the livestock is, but of how effectively the resources can be utilised to meet the requirement of local farmers. Micro livestock and traditional large livestock complement each other because both have unique physical, physiological, behavioural and economic characteristics. Micro cattle are generally reared and valued by farmers because they are climate resilient, have simple requirements, can grow rapidly, can calve easily, show good maternal ability and produce better quality meat and milk. Micro cattle are adapted for use in small fields, terraces and paddy fields. Small hill cattle in the Himalayan regions can negotiate steep slopes and narrow terraces. Micro cattle can integrate in different forms of husbandry easily. Micro cattle suffer less than larger breeds in humid and hot conditions because of their greater ratio of skin area to body mass for shedding heat easily [1–6]. Micro cattle are tolerant to different diseases such as trypanosomiasis in Africa [7]. The adaptability and robustness of micro livestock need attention for their greater use and conservation, their unusual or special characteristic should be noted and genetic origin of breed should be established.

Small animals like cattle, goat, sheep and pig can adapt well to harsh environment conditions such as temperature fluctuations and arid and extreme humidity; thus they are a better choice than large animals. They are distributed in tropical countries and reared for subsistence farming [8–18]. Mini cows usually weigh about 100–150 kg, and sheep and goat weigh about less than 10–12 kg at 12 months of age. The dwarf breeds of cattle, sheep and goats are showing tolerance to trypanosomes and parasites. Poultry birds are also integral part of third-world villages and easily managed by households. Goat, chicken and guinea fowl are reared in backyards in harsh- and disease-prone environment [19]. Turkeys and pigeons are not provided any care and have to scavenge for their sustenance to ecological extremes [20]. Some other birds like guinea fowl which are domesticated in Africa are reared for meat and egg [21–23]. Turkey in Mexico is a hardy and disease-resistant bird [20]. Rabbit is a small animal which has a potential for meat production in poverty-stricken area of the world. Rabbit rearing is popular in different countries [24–26]. Guinea pig is also domesticated for meat in some countries like Bolivia, Peru, Ecuador and India (Table 7.1).

7.2 Goat Farming as Special Boon for Zero Hunger

Goat as species has become important due to concerns over sustainable development and for the upliftment of rural community in terms of economic and food security. Goat has wider genetic base as it is distributed over variable geographical

Table 7.1 Micro livestock with their adaptability and production performance

Breed	Habitat	Production performance	Adaptability
Micro cattle			
Dwarf West African Shorthorn	West African costal forest	125–150 kg	Harsh climate, high rainfall and humid condition, resistance to trypanosomiasis
Muture	Nigeria	160–210 kg	Trypanotolerant
N'Dama	West Africa	200–400 kg	Trypanotolerant and rumen productive, high rainfall
Rodope	Southeastern Europe	200–350 kg	Possibly adaptable to the subtropics
Taiwan Black	Taiwan	250–250 kg	Well adapted to poor tropical condition
Kedah-Kelantan	Malaysia	200–250 kg	Thrive in equatorial environment
Sinhala(Dwarf Zebu)	Srilanka	200–250 kg	
Small Zebu	Somalia	160–230 kg	Adoptable in desolate environment
Dwarf Zebu (Mongalla)	Tanzania, Uganda and Kenya	150–250 kg	Trypanotolerant
Mashona	Zimbabwe	200–250 kg	Resistant to disease and parasites
Criollo	Central and South America	200–300 kg	
Chinampo	Baja California, Mexico	200–350 kg	Tolerant of wild desert condition
Florida scrub	Florida, USA	225–350 kg	Resistant to ticks and screwworm
Hill cattle	Nepal	160–200 kg	Adaptable in poor forage condition
Tibetan dwarf	Tibet	Less than 250 kg	Tolerant to poor forge and high altitudes
Yellow cattle	Southwest and south China	220–380 kg	Adaptable in subtropics and tropics. High temperature and humidity
Micro goats			
American pygmy	United States	15–25 kg	
Sudanese nubian	Northern Sudan	25–30 kg	Arid and semi-arid climate
Sudanese dwarf	Southern Sudan	11–25 kg	
Small East African	Kenya, Uganda, Tanzania	20–30 kg	
Mauritian	Mauritian	25–30 kg	
Criollo	Latin America		
Creole	Caribbean	20–25 kg	
Crioulo	Brazil	30–35 kg	Adapted in harsh environments
Chapper	Pakistan	20–24 kg	Dry regions

(continued)

Table 7.1 (continued)

Breed	Habitat	Production performance	Adaptability
Barbari	India, Pakistan	20–25 kg	
Gaddi (white Himalaya)	Hill districts of northern India	25–30 kg	
Changthangi	Kashmir, India	20 kg	Adaptable in high altitude
Terai	Nepal	8–12 kg	High humidity
Southern hill goat	Nepal	12–16 kg	
Black Bengal	Eastern India and Pakistan	10–14 kg	Humid
Katjang	Southeast Asia, China and Pacific Island	Less than 20 kg	Adaptable in humid condition
Chinese Dwarf	China	20–40 kg	Adaptable in humid tropics
Heuk Yumso	Korea	25–35 kg	
Hajazi	Middle East	20 kg	Harsh desert condition
Sinai	Sinai, Egypt and Negev Desert, Israel	20–50 kg	Drought tolerance
Micro sheep			
West African dwarf	Senegal to Nigeria and South to Angola	25–35 kg	Well adapted in warm and humid condition
Lendim	East and Central Africa	23–40 kg	
Berber	Atlas Mountains	25–41 kg	
Arab	North Africa	40–50 kg	Resistant to extremes of temperature
Southern Sudan dwarf	Eastern and Southern Africa	15–25 kg	
Hejazi	Desert of Arabia	32 kg	Highly acclimatized to drought and privation
Zel	Northern Iran	30–32 kg	Well adapted to subtropical region
Greek Zackel	Mountain and Iceland types	30–40 kg	Resistant to extremes of climate and disease
Sitia	Crete	25–30 kg	Adapted to poor pasturage and extensive herding
Common Albanian		25–35 kg	Marshy area
Zeta Yellow	Yugoslovia	25–35 kg	
Pag	Yugoslovia	20–30 kg	Adapted to scant vegetation and rocky terrain
Corsican	Corsican(France)	25–30 kg	
Galican	Spain	18–25 kg	Adapted to poor pasture
Soay	Scotland	25–30 kg	Adapted to wide temperature variation
Criollo	Latin America	45–70 kg	
Virgin Iceland	Caribbean	35–55 kg	Well adapted to warm humid conditions

(continued)

Table 7.1 (continued)

Breed	Habitat	Production performance	Adaptability
Magra	Northwest India, Pakistan	20–25 kg	Adapted to hot, dry areas
Mandya	Southwest India	25–35 kg	
Hu (Huyang, Lake sheep)	China	35–45 kg	
Javanese Thin-Tail	Indonesia	25–40 kg	
Micro pig			
Chinese Dwarf	China	Less than 70 kg	Adapted to tropical and subtropical condition
Cirollo	Latin America		
Ossabaw	United states	20–30 kg	Adapted to environmental extremes
Kunekune	New Zealand	40–50 kg	

and climatic conditions with differential production potentials. Due to its wider genetic variability, goats can be manipulated to develop suitable adaptable animal in relation to different ecotypes. From a commercial point of view, goats are classified according to their sizes. Small goat breeds weigh about 15 kg in a year. These types of breeds are known as micro goats. Large goat breeds weigh more than 25 kg at 12 months of age, sometimes weighing more than 60 kg. Goats mainly produce meat, and goat meat is widely consumed in different countries, as the main source of protein. Goat's milk is a viable nutrient-rich alternative for cow's milk since it contains lower fat and lactose (milk sugar), is easily digestible in the gut, and is far less inflammatory. Goats also produce the finest and most valuable fibres and durable and fine-textured leather. Goat's manure also has high commercial value.

Goats spend much time in browsing, whereas sheep and cattle in grazing grass. Goats are most efficient and economical for smallholders in all types of production systems. These animals grow faster and can reproduce early in life even under harsh climatic conditions. Goat rearing in the natural ecosystem is an efficient and sustainable method to produce high-quality protein with minimal environmental impacts. Because goats can adapt well to rapidly changing environmental conditions caused by climate change, goat rearing is a profitable option for small farmers, gaining more income as well as generating more employment. Keeping in view the emerging challenges such as climate change, demand for biofuel and growing pressure on land and water resources, goat rearing is profitable and will be least affected by these factors. More importantly, goat rearing is economically viable in all types of production systems and is therefore a viable option for the smallholder to rear it and multiply for commercial use. Goat rearing will enhance the farmer's income and provide more employment to women as it is profitable and less vulnerable to shocks with respect to feed requirement and their management. Goats will play a major role in poverty reduction by providing a better source of income and nutrition to the poor people in the world. Therefore, goats will play a significant role in future animal production as they are profitable, sustainable and well managed by farmers. Goats have the potential for delivering a sustainable increase in income and providing safe, high-quality and affordable livestock products.

Goats have fulfilled their agricultural, economic, cultural and religious roles in all the parts of the world. Goats are the most adaptable and geographically widespread livestock species and one of the first animals to be domesticated by humans around 10,000 years ago at the dawn of the Neolithic period in the Fertile Crescent [27, 28]. The genus *Capra* has contributed to the gene pool of domestic goats, while some studies have suggested that an independent domestication in Pakistan gave rise to the Cashmere breeds [28]. It has been suggested that Mediterranean breeds derive some of their distinct characteristics from the extinct *Capra prisca*. Similarly, the markhor and bezoar are the progenitor of most Indian and Central Asian breeds. The majority of the goat breeds are multipurpose animals [4] and provide meat, milk, hair and hide. The breeds have been classified as meat, milk or fibre type. The dwarf goat breeds of Asia and Africa are specifically known for their meat quality. Goat's meat are selected based on their growth rate, feed conversion efficiency and carcass quality. The most prominent meat, dairy and fibre type goats are described, and its specific use has been summarised. Goat breeds in the tropics are summarised based on their origin, adaptability and speciality in the production environment (Table 7.2).

Goat breeds distributed in extreme climate in different countries are presented in Table 7.2. The goat population and its growth over the last two decades in hunger-affected countries are presented in Table 7.3. In Asia, goat rearing exhibits a positive population growth from 8.376% to 513.333% in 1997. However, some Asian countries like Sri Lanka, the Philippines, Saudi Arabia and Iran and also countries South American countries like Chile, Ecuador and Surinam are experiencing a negative population growth in the last 20 years. In North America, Trinidad was showing a negative population growth in the year 1997. In East Africa, Mauritius is showing a negative population growth during the last 20 years. The West African region was showing a positive population growth over the last two decades. In South Africa, Botswana was showing a −42.639% population growth during 1997–2014. In Central Africa, negative population growths of 10.68% and 13.84% were observed in Congo Republic and Sudan, respectively. In the European region, countries like Bosnia, Croatia, Lithuania, Ukraine, Bulgaria and Albania were exhibiting negative population growth ranging from −18.34% to 65.913% during the last 20 years.

7.3 Integrated Goat Farming

Climatic changes have tremendous impact on agricultural crops and create an alarming situation in food security. Thus, having a stable income from agricultural crops is a big challenge. In order to overcome this problem, the only alternative option is left over agro-based industries like livestock farming. In today's environment, increasing demand for meat in rural and urban areas can help small- and part-time farmers earn good profits by trading with the goat farming. Due to minimum

Table 7.2 Goat breeds with potential for genetic improvement in harsh climatic condition

Breed	Country of origin	Climate & adaptability
Bugri	Pakistan	Dry
Damani	Pakistan	Dry
Fijian	Fuji	Humid
Kaghani	Pakistan	Dry
Kail	Pakistan	Dry
Katjang	Indonesia	Humid
Kacang	Indonesia	Humid
Katukachchiya	Sri – Lanka	Dry
Lehri	Pakistan	Dry
Sudan Desert	Sudan	Very Dry
Tapri	Pakistan	Dry
Terai	Nepal	Humid
Black Dedouin	Egypt	Very Dry
Dera din Panah	Pakistan	Dry
Kamori	Pakistan	Dry
West African dwarf	West – Africa	Humid
Maradi (Red sokoto)	Nigeria	Humid
Mubende	Uganda	Humid
Changthangi	India	Dry Cold
Singhala	Nepal	Cold
The Small East African Goat	Kenya	Dry
Zhongwei	China	
Galla goats	Kenya, Somalia	Dry
Angora Goat	Turkey	Dry cold
Abaza	Turkey	Dry cold
Kalahari Red	South Africa	Semi-arid and arid
Boer	South Africa	Dry
Nigerian Dwarf	West Africa	Dry
Chappar	Baluchistan (Pakistan)	Dry
Creole	West Indies	Dry
Criollo	Mexico	Dry
Kambing, Katajang or pea	Malaysia and Indonesia	Dry
Nepalese Hill	Nepal	Dry
Savana	South Africa	Dry
Syrian Mountain	Syria	Dry
West African Long Legged	Senegal	Dry
Damascus	Syria and Lebanon	Dry
Kurdi	Kurdistan, Iraq, Iran	Dry
Barbari	India	Semi-arid and arid
Black Bengal	India	Hot Humid
Black Bengal	Bangladesh	Hot Humid
Kutchi	India	Dry hot
Ganjam	India	Hot humid
Khasi (Assam local)	India	Humid

(continued)

Table 7.2 (continued)

Breed	Country of origin	Climate & adaptability
Kheri	Nepal	Humid
Assam hill	India	Mountained and high region
Chyangra	Nepal	High mountain, cold region
Marwari	India	Dry
Osmanabadi	India	Semi-arid
Sirohi	India	Dry
Sangamneri	India	Semi-arid
Beetal	India	Arid
Kalahari red	Lisotho	Semi-arid and arid region
Damascus	Syria and Lebanon	Arid
Jamunapari	India	Dry
Jakhrana	India	Semi-arid
Malabari	India	Humid
Surti	India	Semi-arid and humid
Zairibi	Egypt	Dry
Russian White	Russian federation	
Caninde	Brazil	
Repartida	Brazil	
Kottukachchiya	Sri Lanka	Dry

Table 7.3 Goat population and growth in hunger affected countries

Asia	Goat Population (1997)	Goat Population (2014)	Per cent growth
Kuwait	116640	153391	31.508
Turkey	8951000	10347159	15.598
Kazakhstan	679000	2362824	247.986
Jordan	781976	929000	18.802
Saudi Arabia	4433661	3450000	−22.186
China	123158000	187869000	52.543
Iran	26000000	22120000	−14.923
Armenia	21100	30500	44.550
Georgia	54200	60800	12.177
Lebanon	496714	560000	12.741
Kyrgyz Republic	171000	910246	432.308
Azerbaijan	273800	658107	140.360
Thailand	125262	447546	257.288
Oman	925000	2100000	127.027
Turkmenistan	375000	2300000	513.333
Uzbekistan	860000	3181300	269.919
Mongolia	9134800	22008896	140.935
Malaysia	241649	455737	88.595
Philippines	6300000	3690789	−41.416
Indonesia	14162547	19216300	35.684
Nepal	5921960	10177531	71.861

(continued)

Table 7.3 (continued)

Asia	Goat Population (1997)	Goat Population (2014)	Per cent growth
Myanmar	1274950	5782000	353.508
Iraq	1465773	1700000	15.980
Cambodia			
Sri Lanka	520700	299050	−42.568
Bangladesh	34478000	55900000	62.132
North Korea			
Tajikistan	687600	1825183	165.443
India	122721000	133000000	8.376
Pakistan	42650000	66600000	56.155
Afghanistan			
Yemen	6314000	9380000	48.559
Viet Nam	515000	1600275	210.733
Europe			
Belarus	59000	68000	15.254
Bosnia & Herzegovina	170000	74000	−56.471
Croatia	99544	61000	−38.721
Estonia	1700	4900	188.235
Latvia	8400	12600	50.000
Lithuania	16900	13800	−18.343
Montenegro			
Slovak Republic			
Ukraine	853900	668500	−21.712
Romania	654400	1312967	100.637
Macedonia, FYR			
Bulgaria	848742	289308	−65.913
Russian Federation	2445400	2090609	−14.509
Serbia	138604	218603	57.718
Moldova	96000	135500	41.146
Albania	1148000	904214	−21.236
South America			
Chile	738183	450000	−39.040
Uruguay	14800	17400	17.568
Argentina	3428000	4400000	28.355
Brazil	7968169	8851879	11.091
Colombia	1006400	913377	−9.243
Peru	2047720	1910708	−6.691
Suriname	7590	4058	−46.535
Paraguay	122600	143657	17.175
Venezuela	1132656	1432899	26.508
Guyana	79000	82100	3.924
Ecuador	309300	22411	−92.754
Bolivia	1613273	2431041	50.690

(continued)

Table 7.3 (continued)

Asia	Goat Population (1997)	Goat Population (2014)	Per cent growth
North America			
Cuba	138538	608709	339.381
Costa Rica	1600	5160	222.500
Mexico	9208000	8687814	−5.649
Trinidad & Tobago	58250	6854	−88.233
Jamaica	440000	520000	18.182
Nicaragua	6400	7450	16.406
Haiti	1444500	1975000	36.726
Dominican Republic	210000	236000	12.381
North Africa			
Tunisia	1260670	1248200	−0.989
Morocco	4789900	6147200	28.337
East Africa			
Mauritius	91000	26558	−70.815
Kenya	10856261	25430058	134.243
Malawi	1566514	5500000	251.098
Tanzania			
Mozambique	4100000	4780644	16.601
Djibouti	510000	515000	0.980
Rwanda	526253	2725000	417.812
Uganda	853900	14011000	1540.824
Ethiopia	8400000	29112963	246.583
Madagascar	1300000	1473000	13.308
Egypt	3187214	4258000	33.596
West Africa			
Ghana	2634020	6044000	129.459
Senegal	3578000	5381312	50.400
Togo	1229435	2800000	127.747
Gambia	158997	359835	126.316
Mauritania	4200000	5680000	35.238
Nigeria	35000000	71000000	102.857
Côte d'Ivoire			
Burkina Faso	8259311	13891000	68.186
Guinea	825529	1830000	121.676
Mali	6617734	20083130	203.474
Guinea-Bissau	305000	750000	145.902
Niger	8211693	14883559	81.248
Liberia	220000	345000	56.818
Sierra Leone	190000	900000	373.684
Benin	1087000	1755000	61.454
South Africa			
Lesotho	811250	855000	5.393
Botswana	2615000	1500000	−42.639
Namibia	1821009	2240000	23.009
Angola	1720000	4277150	148.672

(continued)

Table 7.3 (continued)

Asia	Goat Population (1997)	Goat Population (2014)	Per cent growth
Zimbabwe	2700000	2750000	1.852
Zambia	700000	2600000	271.429
Central Africa			
Algeria	3121500	5129839	64.339
Gabon	88000	110000	25.000
Cameroon	3700000	4675000	26.351
Congo, Rep.	4612732	4120000	−10.682
Sudan	36037000	31029000	−13.897
Chad	4823550	6850000	42.012
Central African Republic	2213000	5800000	162.088
Central America			
Panama	5200	8455	62.596
El Salvador	15300	15150	−0.980
Honduras	29000	25000	−13.793
Guatemala	109700	113400	3.373
Oceania/Australia			
Fiji	230275	252000	9.434
Timor-Leste	1229435	180000	−85.359

expense and effort, goat farming can be integrated with agricultural crops. Livestock plays an important role in global rural economy. About 20.5 million rural people of developing countries and least developing countries depend on livestock for their livelihood. Livestock systems have both positive and negative effects on the natural resource base, public health, social equality and economic growth. At present, livestock is one of the faster growing agriculture subsectors in developing countries.

Integrating goat farming with agri-silviculture, hortipasture and boundary plantations can be established in wastelands to increase farm income. Composting and vermicomposting are two best alternative options for solid waste biomass management. The goat-fish integrated farming system makes use of livestock waste and conversion of livestock waste into valuable fish protein. The production of manure from goats is around 1.5–2.0 tons per year. It is a good source of organic matter and contains 2.7% nitrogen (N), 1.78% phosphorus (P_2O_5) and 2.88% potassium (K_2O) (Table 7.4) [29]. Manuring helps in the formation of plankton bloom and also acts as a feed material for fish. Vermicomposting is a mesophilic bio-oxidation process in which organic wastes are being catalysed by the dual action of earthworm and microorganism. It is a quicker process of decomposition with the expense of low C/N ratio, resulting in homogenous end product for use as bio-fertiliser.

7.3.1 Integrated Farming for Goat Products

Goat produces meat, milk, fibre and skin. At present, the goat system can also be used as animal bioreactor for the production of therapeutic proteins of high value.

Table 7.4 Nutrient contents (%) of animal faeces

Source	Organic content	Nitrogen	P_2O_5	K_2O	Protein
Goat	60	2.7	1.78	2.88	17.31
Cow	30	0.7	0.30	0.65	4.38
Horse	60	0.7	0.34	0.53	4.38
Pig	30	1.0	0.75	0.85	6.25
Rabbit	50	2.0	1.33	1.20	12.50
Sheep	60	2.0	0.54	1.54	12.54
Chicken	50	1.6	1.25	1.25	10.00

Source: Gaddie and Douglas [29]

The genetically engineered goat can be simultaneously used for biopharmaceutical product development and commercialization of meat, milk and other goat products. This model ensures introducing and producing more varieties of products including therapeutic proteins of high value in the market on the basis of consumer's demand, both in quality and quantity. Farmers raising transgenic goats for therapeutic milk protein must follow the standard supply chain of milk to the processing unit. The farmers working in this chain can earn the money for profitable goat farming. An effective model is proposed for planning integrated goat (Fig. 7.1).

7.3.2 Biosecurity Measures for Goat Farming

Biosecurity is a system of practices adopted in goat farming that prevents infectious diseases from being carried into herd or to premises. It is necessary to develop management programme to prevent the transmission of disease-causing agents to animals by direct or indirect means. Effective biosecurity management practices prevent the spread of disease by minimising movement of biological organisms and their vectors into and within the premises. These approaches help in cost-effective health management, consumers' confidence on farm products and better animal welfare issues.

Protecting animals from infection is the main goal of biosecurity measures. The immunity of animals depend upon host factors (immune status), environmental factors (temperature, stocking rate, pasture condition, etc.) and the disease agent itself. The risk of infection increases due to:

- Contact with other livestock
- Management practice
- Incoming/returning animals
- Vehicle/equipment/traffic flow around the farm
- Feed and water sources
- Manure management
- Movement of other animals as dogs
- Cats, wildlife, rodents and farm visitors

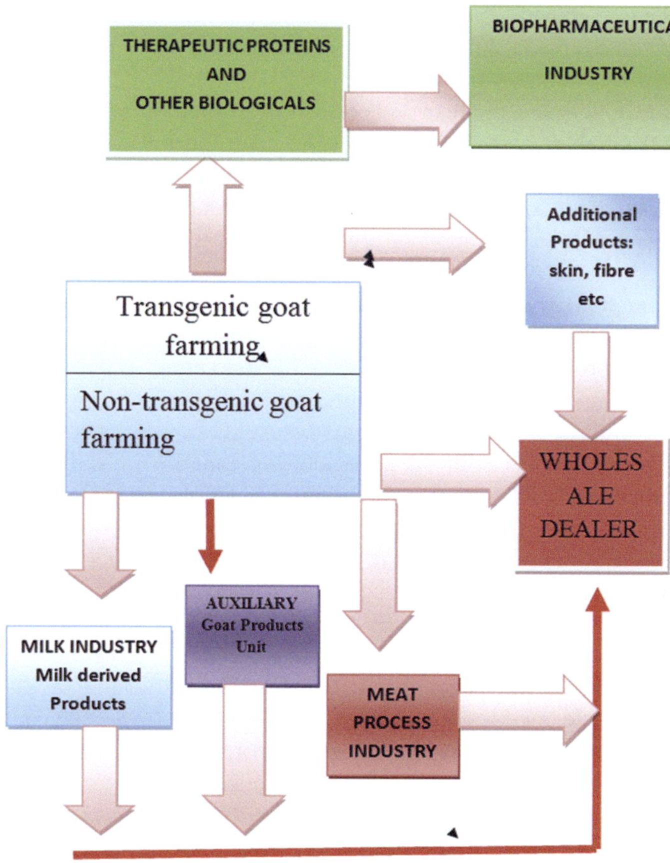

Fig. 7.1 Integrated goat farming for the production of Milk, meat, therapeutic proteins and other additional commercial products. (Source: Goat production and supply chain management)

Therefore, it is essential to follow region-specific suitable package of practices for effective biosecurity measures.

7.4 Goat Farming for Alternative Medicine

The development of therapeutic proteins technology from microbes to mammary gland bioreactor has proven to be a better option for human protein expression. Interestingly, it has been observed that goat's milk contains several types of human proteins. Thus, especially genetically modified goat can be used for the production of therapeutic recombinant proteins due to the production of adequate milk, at the cost of low budget. It is mainly due to goats need small ruminant farmhouse with minimum maintenance expenditure. The state-of-art production of transgenic goat has diverted attention of pharmaceutical industrialists for therapeutic protein production. The genetically engineered goat is supposed to be a better alternate as small ruminant to be used as bioreactor for the production of recombinant therapeutic proteins in their mammary gland. The first human biological drug (antithrombin-AT) was approved by the European Agency of the Evaluation of Medical Products (EAEMP) [30] and then by the US Food and Drug Administration (FDA) [31]. Goat produces about seven- to eightfold increase in the amount of serum than rabbit does. Besides, the goat serum contains about 20 mg/ml of total IgG, which is about two to three times more than the rabbit serum. Thus, goat model bioreactor is a more cost-effective alternative for producing large amount of antibody in a highly commercialised competitive market. The most important fact is getting an antibody from a single animal rather than several kinds of animals.

Presently, the problem related to goat's milk marketing is due to inadequate sequence of commercial process technologies involved in logistic process. Generally, raw milk is thoroughly inspected before its transfer to logistic systems for distribution. So, it is necessary to have additional marketing skills. While marketing fluid goat's milk, the dairy owner should have proper strategy for getting additional income from the kid. Generally, kids are raised to different market weights in different market seasons. At present, powdered goat's milk is gaining commercial importance due to its long shelf-life without any substantial changes in quality [32, 33]. In addition, powdered goat's milk is easy to handle by logistic system for distribution at the cost of low expenditure [34, 35]. The dehydration technology is helpful for the consumers with year-round marketable product. Spray drying is a common technology for making powdered milk. Powdered caprine milk is used for a variety of products like ice cream, chocolate, yogurt, infant formula and cheese [36]. The main reason for the increase demand of goat's milk is for the manufacture of low lactose or lactose-free products. These products are vital for the increasing number of people having lactose intolerance. Goat's milk infant formula is also known as breast milk formula as it has similar prebiotic properties to breast milk. It is a formulated product with the combination of goat's milk as raw material. This market is to fulfil the requirement of bottle-feeding or cup-feeding babies allergic to cow's milk. Dairy goat co-operative (DGC) ranks the first in terms of revenue share in the global market of goat's milk infant formula having 9.99% of market share in 2016 followed by Ausnutria Dairy with a market share of 6.75%.

7.5 Goat Farming-Based Home Industry

The acceptance of goat meat is increasing by non-ethnic consumers. Goat meat has intense flavour, toughness and goaty odour [37]. Its major nutrients are protein, essential amino acids, fat-soluble vitamins, water-soluble vitamins, iron and phosphorus. The goat muscle is also a source of conjugated linoleic acid which plays a role in immune response regulation and energy partitioning and has other health benefits. Chevon has higher thiamine and riboflavin than beef, lamb and veal [38]. Moreover, it has more desirable fatty acids and polyunsaturated to saturated fat ratios than beef and lamb [39]. Castrated goats had lower percentages of stearic, linoleic and total saturated fatty acids [40].

7.5.1 Goat Meat Products

- Goat meat pickle: Goat meat pickle is a shelf-stable value-added product prepared using pre-cooked spent meat.
- Goat meat nuggets: Nuggets are protein- and fibre-rich meat product that needs to be stored under refrigeration.
- Goat meat sausage: It is also a processed meat product that can be stored under refrigeration.
- Goat meat patties: Patties are the most preferred goat meat product among consumers that can be stored under refrigeration.

7.5.2 Milk Product

Goat milk is unique with respect to its digestibility, distinct alkalinity, high-buffering capacity and therapeutic value in human nutrition and medicine [41]. Goat milk is consumed by people allergic to cow's milk and a naturally homogenised product due to its smaller fat globules [42]. Goat milk has more oligosaccharides than cows' milk, with an amount similar to human milk. These act as prebiotics in the gut and may help in maintaining healthy digestive tract by encouraging the growth of beneficial gut bacteria and preventing the growth of harmful bacteria. Goat milk contains slightly less lactose than cows' milk. There has been considerable interest in milk proteins due to availability of bioactive peptides with significant functional properties.

- *Beverage milk:* Fluid milk includes skim or low-fat milk, whole milk and cultured milk products and differs depending on the method of processing.
- *Cheese:* There are about 800 varieties of cheese, and the United States has 400 varieties of goat cheese in the list [44]. Goat cheeses are made from raw and pasteurised goat milk.
- *Buttermilk, yogurt and sour cream:* Buttermilk is usually made from skim milk (less than 0.5% fat) using the by-product from churning butter out of sour cream. Yogurt is made from whole milk (3.25% fat), low-fat milk (0.5–2.5% fat) or skim milk.
- *Kefir:* Kefir is an acidic, slightly foamy, product made from pasteurised and fat-standardised or decreamed (partially skimmed) goat milk
- *Frozen products:* Ice cream and frozen yogurt are also manufactured from goat milk.
- *Evaporated and powdered products:* Powdered products available include whole milk, skim milk, whey and infant foods formula.
- *Other products:* Ghee is also prepared from goat milk. Additional goat milk products made in India include chhana, khoa and paneer cheese. Chhana is an acid- and heat-coagulated milk product. Khoa is a heat-desiccated indigenous milk product used for various sweets. Cajeta in Mexico and Gjetost in Norway are also popular caramelised sweet products made from goat milk. Oggtt is

basically a hard cheese-like product that is considered a very stable and safe dried fermented milk product which is popular in Bedouin areas of Saudi Arabia and Arabian Gulf countries. It is mainly produced by Bedouins during the spring season when milk is produced in excess amounts. Tablets (chewable), whey protein concentrate and athletic supplements are also prepared from goat milk.

7.5.3 Goat Milk Soap

Soap made from goat milk contains natural emollients, vitamins and minerals. Goat milk contains alpha-hydroxyl acids such as lactic acid which help remove dead skin cells from the skin's surface.

7.5.4 Goat Skin Products

Goat skin leather is also known as Morocco leather, and it is known for its softness and lightweight features. It is also one of the most resilient leathers found and is naturally water resistant. Goat skin is more durable than cow skin. It is softer because of the presence of lanolin in the leather. Because of its great qualities such as being very strong and hard-wearing, it is practically used in making leather bags.

7.5.5 Goat Fibre Products

Pashmina, known for its use in beautiful shawls and other handmade items, is produced in Kashmir (India). In the early nineteenth century, cashmere shawls reached their greatest popularity, and the shawls of England, France and the town of Paisley, Scotland, were made to imitate the original Kashmir shawls.

Cashmere: Cashmere is a fibre in demand for its soft, warm and long-wearing characteristics. The cashmere goat wool has a protective outer coat of coarse fibre that is 4–20 cm (1.5–8 in.) in length. The typical range is 16–19 μ. The undercoat is made up of fine, soft fibre commonly called cashmere, which ranges from 2.5 to 9 cm long. It is from the undercoat and is combed off from the goat. White, brown or grey solid coloured goats are preferred over mixed coloured goats. The coarse and down hairs are separated by a mechanical process called dehairing. The long fibres are used in knitted garments. Shorter cashmere fibres go into woven fabrics. The fibre diameter must be less than 19 μ to be classified as cashmere. Traditionally, cashmere fleece has been sold raw in the world market with the end product ranging from dehaired cashmere, woollen or worsted yarns, cashmere blend yarns, woven or worsted fabrics or classic cashmere sweaters and high-fashion garments.

Mohair: Mohair is a lustrous, long fibre renowned for its ability to be dyed into brilliant colours and when blended with other fibre adds sheen to the fabric. These characteristics are sought after by top designers of clothing and soft furnishings worldwide. Mohair is composed mostly of keratin, a protein found in the hair, wool, horns and skin of all mammals. Mohair fibre is approximately 20–40 microns in diameter. Fine hair from younger animals is used for fine applications such as clothing, and the thicker hair from older animals is more often used for rugs, curtaining, upholstery and heavy fabrics intended for outerwear.

7.6 Indicators for Sustainability of Goat Production System

Goats are the best animal component in arid, dry land and rain-fed agro-climatic zones, providing livelihood to millions of households in tropical countries. Goat keeping has been a traditional activity and a primary source of livelihood for smallholders. Goats are distributed in all types of extreme climate and well adapted to that specific environment. They have comparatively low susceptibility to prevailing diseases and can be easily raised with the locally available feed and fodder resources or under intensive, semi-intensive grazing system. Goat is a champion animal for social justice. It provides a potential pathway from poverty to prosperity for resource-poor farmers globally. Modern scientific inputs in goat rearing have tremendous potential for social transformation in most disadvantaged places. Goat production system fulfils all the indicators for sustainability as the system is ecologically stable, economically viable and energy efficient.

Goat as species has become important due to concerns over sustainable development and for the upliftment of rural mass in terms of economic and nutritional stability. Goat has wider genetic base as it is distributed over varied geographical and climatic conditions with differential production capabilities. Due to its wider genetic variability, goats can be manipulated to develop suitable adaptable animal in relation to different tropical regions. Goats can be reared along with other livestock without any serious competition due to preferential grazing habit. Goats spend more time in browsing, whereas sheep and cattle in grazing grass. Goats are most efficient and economical for smallholders in all types of production systems. These animals have faster growth rate and reproduce at an early age under harsh climatic conditions. The tropical region will be highly affected by the climate change, and goats in that region will play a significant role in providing livelihood to farmers. Goats have wide adaptability and genotypes are observed in all extreme environmental conditions. Therefore the genotypes can be utilised for better use during varied climatic conditions. Goats are of special economic importance in arid, in semiarid and in dry land agriculture, where other livestock farming is not economically feasible.

Ecological stability of the farming system should be the foremost criterion for sustainable development. Goats can thrive on many foodstuffs which are not utilised by any other species of livestock. Goats are able to adjust well in hot, humid and waterlogging condition efficiently than other ruminants. Goats integrate well in mixed agriculture as they consume leafy wastes and clear weeds and provide manure. They can graze in rough terrain, desert field for longer periods without water and utilise the roughage not used by anyone.

Goats generally have a long snout and upright tail. The mouth is unusual in having a mobile upper lip and grasping tongue, which permits the animal to nibble tiny leaves from thorny trees. However, goat among all the domestic animals has been labelled as the enemy of ecology with regard to grazing habit. But this concept is not valid anymore as there are several scientific works to disprove it. Studies at different institutes indicated that small ruminants do not play a major role in ecological destruction and their role in ecological degradation has been exaggerated [43]. Quite contrarily, some experts opine that goats improve the grazing land by dispensing seeds, and they have been widely recognised as helpers in controlling many weeds such as *Nesetta trichaforma, Rubus fruticosus and Rosa rubiginosa*. In countries like France, Italy, Australia and Syria, there is no official bias against goats. There is no correlation between goat density and deforestation. Higher concentrations of goats are found in semiarid and rain-fed region in the tropics, and the area will be targeted for future agricultural exploitation.

Goat farming system is generally capable of fulfilling most of the basic needs and providing work for all local inhabitants with the guarantee of their satisfactory living standard. Goats are generally an adjunct to cropping system where they make use of natural vegetation and crop residues supplemented by tree leaves. Goat farming is remunerative among all the livestock farming systems in the entire farm situation. As it is a low investment proposition for farmers, it is a very assured way of income to marginal and small farmers who survive with part-time employment as agricultural labour. The small size is especially significant as it relates directly to economic, managerial and biological advantages; this is especially true in marginal and rangeland areas where rainfall and available feeds are sparse.

It has been reported that a goat keeper could earn about 250 (USD 4.17), 475 (USD 7.92) and 800 (USD 13.33) Indian rupees per month having small-sized, medium-sized and large-sized flocks, respectively [44]. The economics of 1 buffalo vs. 5 goat unit in Rajasthan (India) was studied, and it was observed that the profit per annum was 1945 Indian rupees (USD 32.42) from goat unit and 1755 Indian rupees (USD 29.25) from buffalo unit [45]. Similarly a study of drought-prone total gross income of small and marginal farmers was accounted from small ruminants [46].

The average gross income per household was highest in goat farming as compared to other livestock and crop farming system. In arid region, the goats were more economical than sheep and cattle. Goats in desert kept under range conditions are 40–160% more economical than sheep and that an indigenous goat may provide 250 Indian rupees (USD 4.17) per year to a small farmer besides 2 quintal of manure and benefits of clearing obnoxious weeds and thorny bushes [46].

Energy efficiency of the system is measured as output-input energy ratios. Therefore, it is necessary to understand the ecosystem and the farmer's indigenous practice and knowledge to design highly energy-efficient system. Inputs such as support energy, land and labour are not that costly for the goat production system. As far as the grazing habit is concerned, goat can thrive on vegetation, which is not suitable for other animal species. As a ruminant, it can use roughages and nonconventional feeds for its energy utilisation and growth. Land mainly targeted for goat grazing is rangeland and wasteland. Another major input is labour, which is the lowest one in goat farming in tropical countries [47]. Goat rearing is mainly handled by women and children in village conditions. Moreover, the output of the system is

Table 7.5 Partial enterprise budget for subsistence goat production

Description:	
Gains (comparative additional gains)	
Costs saved	
Extra revenue	
	Food security
	Women prestige
Losses (comparative additional losses)	
Extra costs	Payment for buck service
	Extra labour for animal care
	Extra tree fodder for buck
	Improved shelter for buck
Loss of revenue	Use fodder for fire/timber income from selling corn used for extra feeding
Extra profit	Total additional gains – Total additional losses

Source: Andreasen and Qwist-Hoffmann [48]

milk, meat, fibre, manure and breeding stock. The meat demand is high, so question of overproduction of animal does not arise in the present situation. However, excreta of goat urine and dung have been proven to be rich sources of nitrogen for soil fertility.

7.7 Subsistence Goat Farming

The goats are raised under subsistence farming in tropical countries. In goat rearing, social values and nutritional security are major factors and can surpass economic value. Economic analysis of different subsistence farming such as pastoral system, arid desert and high humid in small holder rearing is required for improving management system and increasing farmer's income. The components of budget analysis are needed to be determined for each production environment. The first step will be to describe the enterprise with respect to production system and general assumption about flock size, kidding rate, mortality rate, etc. The next step will be to assess the total gains and total losses. The total gains and losses can be prepared following Table 7.5.

References

1. Maule JP (1989) In: Centre for Tropical Veterinary Medicine, Edinburgh University (ed) The cattle of the tropics, Edinburgh
2. Clutton-Brock J (1981) Domesticated animals from early times. British Museum (Natural History), London
3. Epstein H, Mason IL (1984) Cattle. In: Evolution of domesticated animals, Mason IL. Longman, London, pp 6–27
4. Mason IL (1996) A world dictionary of livestock breeds, types and varieties, 4th edn. C.A.B. International, Wallingford

5. Joshi NR, Phillips RW (1953) Zebu cattle of India and Pakistan, FAO Agricultural Study No. 19. Food and Agriculture Organization of the United Nations, Rome, p 297
6. Peilieu C (1984) Livestock breeds of China. FAO Animal Production and Health Paper 46. Food and Agriculture Organization of the United Nations, Rome. 217
7. ILCA (1979) Trypanotolerant livestock in west and central Africa, General Study, vol 1. International Livestock Centre for Africa, Addis Ababa
8. Acharya RM (1982) Sheep and goat breeds of India, FAO Animal Production and Health Paper No. 30. Food and Agriculture Organization of the United Nations, Rome
9. Haenlein GFW, Ace DL (1983) Goat extension handbook. University of Delaware Cooperative Extension Service, Newark
10. Devendra C, McLeroy GL (1983) Goat and sheep production in the tropics, 2nd edn. Longman, London, p 271
11. Devendra C, Burns M (1983) Goat production in the tropics. Commonwealth Agricultural Bureaux, Buckinghamshire
12. Gall C (ed) (1981) Goat production. Academic, London, p 619
13. Carles AB (1983) Sheep production in the tropics. Oxford University Press, London, p 213
14. Husnain HV (1985) Sheep and goat in Pakistan. FAO Animal Production and Health Paper No. 56. Food and Agriculture Organization of the United Nations, Rome
15. Mason IL (1980) Prolific tropical sheep. FAO Animal Production and Health Paper No. 17. Food and Agriculture Organization of the United Nations, Rome
16. Cole HH, Garrett WN (1980) Animal agriculture. W. H. Freeman and company, San Francisco
17. Epstein H, Bichard M (1984) Pigs. In: Mason IL (ed) Evolution of Domesticated Animals. Longman, London, pp 145–162
18. Pond WG, Maner JH (1984) Swine production and nutrition. AVI Publishing Company, Westport
19. Crawford RD (1984) Domestic fowl. In: Mason IL (ed) Evolution of domesticated animals. Longman, London, pp 298–311
20. Berg R, Halvorson D (1985) Turkey management guide. Minnesota Turkey Growers Association, St. Paul
21. Belshaw RHH (1985) Guinea fowl of the world. Nimrod Book Services, Hampshire
22. Blum JC, Guillaume J, Leclercq B (1975) Studies of the energy and protein requirements of the growing guinea-fowl. Br Poult Sci 16:157–168
23. Mongin P, Plouzeau M (1984) Guinea-fowl. In: Mason IL (ed) Evolution of Domesticated Animals. Longman, London, pp 322–324
24. Cheeke PR, Patton NM, Lukefahr SD et al (1986) Rabbit production. 6th (edn) The Interstate, Danville
25. Ministry of Livestock Development (1981) Rabbit production. Kenya Ministry of Livestock Development. Copies can be obtained from the Agricultural Information Centre, P.O. Box 14733, Nairobi
26. FAO (1986) Backyard rabbit rearing: some basic husbandry practices. Small animals for small farms. GAN-16. Food and Agriculture Organization of the United Nations, Regional office for Latin America and Caribbean, Santiago
27. Pringle H (1998) Neolithic agriculture: reading the signs of ancient animal domestication. Science 282:1448
28. Porter V (1996) Goats of the world. Farming Press, Ipswich
29. Gaddie ER, Douglas DE (1975) Earth worm for ecology and profit. Scientific earthworm farming, vol 1. Bookworm Publishing Company, Ontario
30. Schmidt C (2006) Belated approval of first recombinant protein from animal. Nat Biotechnol 24:877
31. Lavine G (2009) FDA approves first biological product derived from transgenic animal. Am J Health-Syst Pharm 66:518
32. Silanikove SSG, Kebede E et al (2010) In: Solaiman SG (ed) Business plan, production enterprise and marketing strategy in: goat science and production. Wiley-Blackwell Publishers, New York, pp 339–357

33. Fonseca CR, Bordin K, Fernandes AM et al (2013) Storage of refrigerated raw goat milk affecting the quality of whole milk powder. J Dairy Sci 96:4716–4724
34. Bylund G (2003) Dairy processing handbook. Tetra Pak Processing Systems AB, Lund, pp 361–374
35. Tehrany E, Sonneveld K (2010) Packaging and the shelf life of milk powders. In: Food packaging and shelf life, a practical guide. CRC Press, Boca Raton, pp 127–141
36. Kajal MFI, Wadud A, Islam MN et al (2012) Evaluation of some chemical parameters of powder milk available in Mymensingh town. J Bangladesh Agric Univ 10(1):95–100
37. Wong E, Johnson CB, Nixon LN (1975) The contribution of 4- methyloctanic (hircinoic) acid to mutton and goat meat flavor. NZ J Agric Res 18:261–266
38. Webb EC, Casey NH, Simela L (2005) Goat meat quality. Small Rumin Res 60:153–166
39. Werdi Pratiwi NM, Murray PJ, Taylor DG et al (2006) Comparison slaughter weight and castration on fatty acid profiles in the longissimus thoracic muscle from male Boer and longissimus thoracic muscle from male Boer and Australian feral goats. Small Rumin Res 64:94–100
40. Park YW, Haenlein GFW (2007) Goat Milk, its products and nutrition. In: Hui YH (ed) Handbook of food products manufacturing. Wiley, New York, pp 447–486
41. Haenlein GFW (2004) Goat milk in human nutrition. Small Rumin Res 51:155–163
42. Sanders GP (1969) Cheese varieties and descriptions; USDA Agric. Handbook no. 54. USDA, Washington, DC
43. RTF (1987) Government of India. Report of the task force to evaluate the impact of sheep and goat rearing in ecologically fragile zones. Government of India, Ministry of Agriculture, Department of Agriculture and Cooperation, New Delhi
44. Singh K, Ram K (1987) Economic analysis of goat keeping in the goat breeding areas of Punjab. Ind J Anim Sci 57:317–323
45. Pasha SA (1990) Sustainability and viability of small marginal farmers with reference to animal husbandry and common property resources. Paper presented in National Seminar held on October 8 to October 10 at NIRD, Hyderabad, India
46. Ghosh PK, Khan MS (1980) The goat in desert environment, Research Bulletin No. 12. CAZRI, Jodhpur
47. Sastry NSR (1995) Livestock sector of India: regional aspects. International Book House, Lucknow
48. Andreasen L, Qwist-Hoffmann H (1997) Inter regional project for participatory upland conservation and development (Field Document 6/97) –NEPAL_A framework for a participatory economic evaluation of improved goat production by women's groups in the Bhusunde Khola watershed

8 World Hunger and Poverty

Hunger is a sense of feeling to intake some edible things, orally, to get relief from discomfort caused due to empty stomach. Hunger is the biological signal when the level of blood sugar is too low. Food and nutrients reach in proteins, carbohydrate and essential elements help to fill up stomach and curb hunger pangs. Biologically, hunger is a sensation that represents the body's requirement for eating some quality food. Two hormones named ghrelin and leptin regulate hunger and satiety in the body. Leptin otherwise known as *satiety hormone* inhibits hunger by signalling the central nervous system. World hunger refers to hunger aggregated to the global level and which can be expressed in terms of food security and malnutrition. The USDA defines food insecurity as a state in which "consistent access to adequate food is limited by a lack of money and other resources at times during the year". In short terms, food insecurities are "struggle *to avoid hunger*", "*hungry or at risk of hunger*" and "hungry *or faced by the threat of hunger*". Malnutrition is a condition resulting from insufficient intake of biologically necessary nutrients. About 795 million people in the world are lacking of adequate food to lead a normal healthy life. It means one in nine people on earth is suffering from hunger. The most of the hunger victimised people live in developing countries or least developing countries. About 12.9% of world population is suffering from undernourished.

Hunger is a resulted state of condition of an individual due to the cause of poverty. Poverty is not having enough material possessions or income for a person's needs. Poverty is a multifaceted concept depending on the social, economic and political scenario of a state. Absolute poverty is the complete lack of the means necessary to meet basic personal need such as food, clothing and shelter.

B. K. Behera et al., *Move Towards Zero Hunger*,
https://doi.org/10.1007/978-981-32-9800-2_8

8.1 Types of Hunger

On the basis of causes like poverty, undernourishment and micronutrient deficiency, the condition of hunger can be categorised into three types. Following are the three major types of hunger noticed in rural areas of developing countries and least developing countries.

8.1.1 Acute Hunger

It is the most extreme form of hunger and is defined as undernourishment over a definable time period. Acute hunger arises due to crises like droughts, El Niño, wars, population displacements and disasters. It severely affects people who are already suffering from chronic hunger. The people from Afghanistan, the Democratic Republic of the Congo, Ethiopia, Nigeria, South Sudan, Sudan, Syria and Yemen face acute hunger. In Africa, 72 million people suffer from acute hunger. More than 113 million people across 53 countries experienced "acute hunger" during 2018 because of wars and climatic disasters. Africa is the worst-hit region and suffers from acute hunger. The key factors which are responsible for hunger were conflict, climate-related natural calamities like drought and floods and economic problems. Acute food insecurity is when a person's inability to consume adequate food puts their lives or livelihoods in immediate danger.

In 2019 the main causes of acute hunger and food insecurity were conflict (74 million people affected), climate (29 million people affected) and economic shocks (10 million people affected). The number of people in the world facing food crises has remained over 100 million in the last 3 years, and the number of countries affected has risen. Moreover, an additional 143 million people in another 42 countries are just one step away from facing acute hunger. The eight countries facing acute hunger are Afghanistan, the Democratic Republic of the Congo, Ethiopia, Nigeria, South Sudan, Sudan, Syria and Yemen.

8.1.2 Chronic Hunger

It defines the state of long-term undernourishment. The body absorbs less food than it needs. Chronic hunger is usually associated with poverty. Chronically hungry people do not have enough resources for healthy nutrition, clean water or healthcare. Chronic hunger is when a person is unable to consume enough food to maintain a normal, active lifestyle over an extended period. The hunger issues are complicated by various natural disasters, and 821 million people in the planet are hungry. Chronic hunger solutions are initially thought can be solved by food distribution. But lately it was realised that the solution to chronic hunger involves people in the process of feeding themselves, either through sustainable food production, increased income earning potential or improved ways to store and market their food produced. Family farming adoption is one of the process by which chronic hunger problem can be solved.

8.1.3 Hidden Hunger

It is a form of chronic hunger. It is the micronutrient deficiency in body. Hidden hunger refers to deficiencies of key micronutrients (vitamins and minerals) in humans. Hidden hunger occurs when the quality of food people eat does not meet their nutrient requirement, so the food is deficiency in micronutrient. The prevalence of micronutrient deficiencies is particularly high in South Asia, despite of better economic growth, agricultural output and healthcare. Hidden hunger not only harms individuals but can inhibit the overall development of an affected region, as the efficiency and health of people decrease. The problem of hidden hunger and chronic hunger can be solved by adopting family farming by people in all the regions.

8.1.3.1 Prevalence of Hidden Hunger

Micronutrients are essential for the growth and development of body, especially for infants. Micronutrients stimulate cellular growth and metabolism. Most commonly, deficiency of iron, iodine and vitamin A is the widespread form of micronutrient malnutrition with public health. Other micronutrients have shown to play a significant role in preventing specific disease conditions (e.g. folic acid and calcium) or in promoting growth (e.g. zinc). Mainly, anaemia and goitre are the two visible indicators for micronutrient deficiency. In developing country like India, 63% of children below 5 years in urban areas and 72% in rural areas are suffering from anaemic problem.

8.1.3.2 How to Overcome with Micronutrient Malnutrition

Various technical application practices like supplementation, fortification and dietary diversification are being suggested to overcome with the problem of micronutrient malnutrition.

Supplementation

Various types of nutrients can be delivered to patient by means of nutraceutical syrup or pills. Under normal circumstances, supplementation programmes are used as a short-term measure and are then replace with long-term, sustainable food-based measures such as fortification and dietary modification, usually by increasing food diversity. In most of the primary healthcare centre, supplementation is commonly practised to provide iron and folic acid to pregnant women and vitamin A to infants and children under 5 years of age.

Food Fortification

As defined by the World Health Organization (WHO) and Food and Agriculture Organization of the United Nations (FAO), fortification refers to "the practice of deliberately increasing the content of an essential micronutrient, i.e. vitamins, and minerals (including trace elements) in a food, so as to improve the nutritional quality of the food supply and to provide a public health benefit with minimal risk to health", whereas enrichment is defined as "synonymous with fortification and refers to the addition of micronutrients to a food which are lost during processing". Thus, it is a process of addition of micronutrient (key vitamins) and minerals such as iron, iodine and zinc to staple foods such as rice, milk and salt to improve their nutritional content. These nutrients may or may not have been originally present in the food before processing. Often, due to the deficiency of certain minerals, the staple foods produced in that geographical location lack many important essential elements and proteins which are needed for the growth and development of body. Addition of micronutrients to staples can prevent large-scale deficiency diseases.

Bio-fortification

Bio-fortification or plant breeding for the specific purpose is a method for enhancing the nutritional value of crop by using better techniques of plant breeding or recombinant technology by incorporating a desire gene. There have been some reported successes, including high-protein maize, high-carotene sweet potato and cassava and rice (golden rice). This is still in the experimental stage but has a potential to remarkably improve the nutritional status of deprived sections.

Advantages of Fortification

This strategy can lead to relatively rapid improvements in the micronutrient status of a population and at a very reasonable cost, especially if advantage can be taken of existing technology and local distribution networks. Food fortification reinforces and supports ongoing nutrition improvement programmes and should be regarded as part of a broader, integrated approach to prevent hidden hunger, thereby complementing other approaches to improve micronutrient status. These additions have helped to solve public health problems, such as salt iodization to prevent goitre. Food fortification will continue to be an important tool, not only to treat or prevent specific nutritional deficiencies but also to promote a general state of well-being in different populations and possibly to prevent certain chronic diseases.

8.2 Initiative Factors for Hunger

Hunger or malnutrition for the society is a vicious cycle. If the mother is undernourished, then the child is undernourished and suffers from malnutrition in the womb. It arrests the physical and mental development of the child. Hunger and malnutrition have many causes such as:

Wars and Conflicts

Armed conflicts force people to leave their native places and not to cultivate their fields or not be able to run their family business. Roads and agricultural infrastructure such as irrigation facilities are destroyed. Due to the limited security, trade also suffers; food becomes scarce and expensive.

Natural Disasters

Droughts and floods destroy agriculture production. Droughts over many consecutive years weaken the resilience of the population. Natural disasters affect the hunger and also affect agricultural production from all angles such as seed supplies and livestock production performance.

Poverty

The poor people have no adequate resources for food, clean water, education and better health coverage.

Inequality

The inequality between rich and poor is increasing. The poor and hungry have almost no chance to free themselves from their hardship. The recently introduced Food Security Standard (FSS) should control food and nutrition security through greater justice worldwide.

Biased Global Trade

The rich states determine the rules of international policy. Developing countries primarily export raw materials, and the profits are enjoyed by rich states.

Poor Governance

The governments should align their policies to fulfil the needs of the poorest population. There should be adequate strategies to promote agriculture at region level to ensure uniform food supply to all. Social cohesiveness and corruption are also contributing factors for poverty and hunger.

Waste of Resources and Climate Change

Expansion of deserts, soil erosion, water scarcity and extreme weather phenomena as a result of climate change are becoming particularly apparent in countries that already suffer from hunger and poverty.

8.3 Global Hunger Index

Hunger usually refers to distress associated with the lack of food. The Food and Agriculture Organization (FAO) defines food deprivation or undernourishment on the consumption of fewer than about 1800 kcal a day. It is a minimum requirement of a person to lead a healthy and productive life [1]. Undernutrition is deficiency of energy, protein or essential vitamins and minerals. Undernutrition is the result of low intake of food with respect to quantity or quality and poor utilisation of nutrients due to ill health. Moreover this is affected by a number of factors such as household food insecurity, inadequate maternal health, inadequate sanitation and health practices and unavailability good drinking water [1]. Malnutrition refers to both food deficiency (undernutrition) and overnutrition (problem of unbalanced diets) with or without low intake of micronutrient-rich foods.

GHI reflects the nutrition situation of the whole population including children; status of dietary energy, protein or micronutrients; and status of physical and cognitive development and death. GHI was first released by the International Food Policy Research Institute and Welthungerhilfe in 2006. The index first included the following three equally weighted nonstandardised indicators:

- The proportion of the population that is undernourished
- The prevalence of underweight in children under 5
- The under-five mortality rate

Then subsequently the formula was revised during 2015 and included child underweight, child wasting and child stunting to construct index by standardising each of the component indicators [2]. Child underweight has been replaced with wasting and child stunting. The other component indicators – undernourishment and child mortality – remained as it is. GHI was not standardised; thus by standardising the value of four component indicators, their effect on GHI scores can be balanced in any given year and over time. The GHI is designed to raise awareness and understanding of regional and country differences in the struggle against hunger. GHI will initiate action to reduce hunger around the world [3]. Hunger has multiple aetiologies and quantifies the effect of each dimension, and describing the index is important one. GHI combines the four component indicators into one index [4]. The indication and weightage for calculating index are described below (Table 8.1).

The Global Hunger Index (GHI) is a tool which measures hunger and track hunger at international, national and regional level. It is essential to understand how the GHI scores are calculated and their significance [5]. GHI scores are calculated annually by the International Food Policy Research Institute (IFPRI) to evaluate progress and setbacks in the fight against hunger. GHI results are calculated using a three-step method that captures the multidimensional nature of hunger using accessible information from different sources. GHI score index was designed as per the hunger rating scale, and the following four indicators are considered for each country.

Undernourishments

The proportion of the undernourished population (i.e. the caloric consumption of which is insufficient)

Kids Wasting

Table 8.1 Feature of the Global Hunger Index

Dimensions	Indicators	Weight	Reasons for inclusion
Inadequate food supply	Undernourishment	1/3	>Measures insufficient food supply, an important indicator of hunger
			>Refers to the entire population, both children and adults
			>Used as a lead indicator for international hunger targets
Child undernutrition	Wasting	1/6	>Goes beyond calorie availability, considers aspects of diet quality and utilisation
	Stunting	1/6	
			>Children are particularly vulnerable to nutritional deficiencies
			>Is sensitive to uneven distribution of food within the household
			>Stunting and wasting are the suggested nutrition indicators for the Sustainable Development Goals (SDGs)
Child mortality	Under-five mortality rate	1/3	>Death is the most serious consequence of hunger, and children are most vulnerable
			>Improves the GHI's ability to reflect micronutrient deficiencies
			>Wasting and stunting only partially capture the mortality risk of undernutrition

The proportion of children under the age of 5 who are wasted (i.e. those with low height and weight reflecting undernutrition)

Kids Stunting

The children under the age of 5 who are stunted (i.e. who have low height for their age, reflecting chronic undernutrition)

Kids Mortality

Mortality rates of kids under 5 years of age (partly reflecting the deadly mixture of insufficient nutrition and unhealthy environments)

Global hunger severity scales are classified as low, moderate, serious, alarming and extremely alarming. The severity scales and effect of hunger level represented below.

GHI severity scale				
Low	Moderate	Serious	Alarming	Extremely alarming
≤9.9	10.0–19.9	20.0–34.9	35.0–49.9	≥50.0

Hunger countries with different sustainable indicators are presented in Table 8.2.

Table 8.2 119 hunger countries with different indicators of sustainable development

Countries with the highest number of undernourished (2016)	Average dietary energy supply (kcal/cap/day)	Prevalence of undernourishment (%)	Prevalence of severe food insecurity (%)	Stunting, children under 5 (%)	Wasting, children under 5 (%)	Safely managed drinking water (% pop. using)	Forest area (% total land area)	Agricultural area (% total land area)	Net emissions in land use (CO_2 eq. Gg)
Asia									
Afghanistan	2024	30.3	16.0	*40.9*	*9.5*		2	58	0
Armenia	2928	4.3	3.8	9.4	4.2	60.6	12	59	−146
Azerbaijan	3131	<2.5		18	3.1	71.5	14	58	−1472
Bangladesh	2487	15.2		36.1	14.3	55.7	11	71	29036
Cambodia	2451	18.5	14.4	32.4	9.6	24.1	53	31	23452
China	3211	8.7		8.1	1.9		22	56	−311371
Georgia	2848	7.4	8.9			73	41	34	5
India	2492	14.8		38.4	21		24	60	122469
Indonesia	2832	7.7		36.4	13.5		50	31	1468854
Iran	3120	4.9				91.4	7	28	67087
Iraq	2491	27.7					2	21	−2038
Jordan	2570	13.5	13.9			93.3			
Kazakhstan	3225	<2.5	1.4	8	3.1		1	80	172
Kuwait	3472	<2.5		4.9	3.1	100	0	8	−15
Kyrgyz Republic	2793	6.5	4.4	12.9	2.8	66.3	3	55	−835
Lebanon	2765	10.9				47.5	13	64	−4
Malaysia	2964	2.9		20.7	11.5	92.1	68	26	−138287
Mongolia	2453	18.7	2.8	22.5	10.8	1	8	71	26597
Myanmar	2711	10.5		29.2	7		44	20	104038
Nepal	2666	9.5	7.8	35.8	9.7	26.8	25	29	6200
North Korea	3318	<2.5	<0.5			98	63	17	−39595

Oman	3126	5.4		14.1	7.5	88.5	0	5	−5
Pakistan	2432	20.5				35.6	2	48	28602
Philippines	2601	13.7	12.9	33.4	7.1		28	42	−60309
Saudi Arabia	3255	5.5					0	81	0
Sri Lanka	2577	10.9		17.3	15.1		33	44	4459
Tajikistan	2213		7.8			47.4	3	34	0
Thailand	2784	9		10.5	5.4		32	43	15333
Turkey	3724	<2.5		9.5	1.7		15	50	−64854
Turkmenistan	2779	5.5		11.5	4.2	86.1	9	72	0
Uzbekistan	2730	7.4					8	63	−16407
Viet Nam	2828	10.8	2.3	24.6	6.4		48	39	−18836
Yemen	2069	34.4		46.5	16.3		1	45	0
Europe									
Albania	3223	5.5	10.5			68.9	28	43	−325
Belarus	3236	<2.5				94.4	43	42	−557
Bosnia and Herzegovina	3221	<2.5	1.5			88.6	43	43	175
Bulgaria	2926	3				96.6	35	46	−9875
Croatia	3059	<2.5	0.8			90.5	34	28	−3836
Estonia	3253	2.8	<0.5			81.7	51	23	4400
Latvia	3174	<2.5	0.7			81.9	54	31	−10843
Lithuania	3417	<2.5	<0.5			91.7	35	47	442
Macedonia, FYR	2957	4.1	3.3			83.5	40	50	8
Moldova	2638		3.4			70	13	74	−1087
Montenegro	3491	<2.5	2.3	9.4	2.8	89.7	61	19	82
Romania	3358	<2.5	4.1			87.8	30	59	−163326
Russian Federation	34.1	<2.5				75.5	50	13	−97605

(continued)

Table 8.2 (continued)

Countries with the highest number of undernourished (2016)	Average dietary energy supply (kcal/cap/day)	Prevalence of undernourishment (%)	Prevalence of severe food insecurity (%)	Stunting, children under 5 (%)	Wasting, children under 5 (%)	Safely managed drinking water (% pop. using)	Forest area (% total land area)	Agricultural area (% total land area)	Net emissions in land use (CO_2 eq. Gg)
Serbia	2728	5.6	2.1			88.1	31	39	−1293
Slovak Republic									
Ukraine	2948	3.3				92.2	17	72	−4645
South America									
Argentina	3250	3.8	8.7			98.5	10	54	100293
Bolivia	2355	19.8		16.1	2		50	35	89914
Brazil	3184	<2.5					59	34	308686
Chile	3032	3.3	4.4	1.8	0.3	98.2	24	21	−105088
Colombia	2977	6.5				71.1	53	40	18233
Ecuador	2614	7.8	7.1	23.9	1.6	74.4	50	22	33899
Guyana	2813	7.5		12	6.4		84	9	17575
Paraguay	2635	11.2		5.6	1		38	55	143319
Peru	2682	8.8		13.1	1	50.2	58	19	72039
Suriname	2780	7.6					98	1	4549
Uruguay	32.6	<2.5					11	83	−10496
Venezuela	2454	11.7					53	24	67448
North America									
Costa Rica	2905	4.4	4.8			89.8	55	34	−11343
Cuba	3594	<2.5					31	60	−13946
Dominican Republic	2697	10.4		7.1	2.4		42	49	−8702
Haiti	2222	45.8					3	67	138
Jamaica	2780	8.9		6.2	3.6		31	41	805

Mexico	3143	3.8	8.9	12.4	1	42.6	34	55	8705
Morocco	3470	3.9				68.8	13	69	−1465
Nicaragua	2689	16.2				58.7	26	42	163
North Africa									
Trinidad and Tobago	3138	4.9					46	11	−498
Tunisia	3345	4.9				92.7	7	65	−291
East Africa									
Djibouti	2575	19.7					0	73	0
Egypt	3537	4.8	10.1	22.3	9.5		0	4	−218
Ethiopia	2330	21.4		38.4	9.9	10.5	13	36	17720
Kenya	2217	24.2	35.6	26.0	4.0		8	49	−30708
Madagascar	1907	43.1					21	71	21743
Malawi	2246	26.3	52.4	37.1	2.7		33	61	4414
Mauritius	3043	5.8	5.9				19	42	−15
Mozambique	2252	30.5					48	64	39199
Rwanda	2203	36.1		37.9	2.2		20	73	733
Tanzania	2241	32.0		34.4	4.5		52	45	220636
Uganda	2033	41.4		28.9	3.6	6.4	10	72	26105
West Africa									
Benin	2699	10.4		34	4.5		38	33	10841
Burkina Faso	2689	21.3	23.8	27.3	7.6		19	44	9288
Côte d'Ivoire	2625	20.7		21.6	6	45.8	33	65	5589
Gambia	2656	9.6	25.9	25	11.1		48	60	−217
Ghana	3048	6.1	7.9	18.8	4.7	26.9	41	69	8360
Guinea	2547	19.7	39.9	32.4	8.1		26	59	14090
Guinea-Bissau	2255	26	34	27.6	6		70	58	1475
Liberia	2204	38.8	56.6	32.1	5.6		43	28	1372
Mali	2956	6		30.4	13.5		4	34	7425

(continued)

Table 8.2 (continued)

Countries with the highest number of undernourished (2016)	Average dietary energy supply (kcal/cap/day)	Prevalence of undernourishment (%)	Prevalence of severe food insecurity (%)	Stunting, children under 5 (%)	Wasting, children under 5 (%)	Safely managed drinking water (% pop. using)	Forest area (% total land area)	Agricultural area (% total land area)	Net emissions in land use (CO_2 eq. Gg)
Mauritania	2814	11.3		27.9	14.8		0	39	−1518
Niger	2581	14.4	37.2	42.2	10.3		1	36	1574
Nigeria	2518	11.5	24.8	43.6	10.8	19.4	7	78	191754
Senegal	2473	11.3		17	7.2		43	46	5204
Sierra Leone	2399	25.5	70.6	37.9	9.4		43	55	5900
Togo	2520	16.2	30.5	27.5	6.7		3	70	6589
South Africa									
Angola	2259	23.9		37.6	4.9		46	47	107837
Botswana	2302	28.5	39.8				19	46	34814
Lesotho	2579	12.8	50	33.2	2.8		2	78	−196
Namibia	2260	25.4	39.8	23.1	7.1		8	47	9188
Zambia	2017	44.5		40	6.3		65	32	344511
Zimbabwe	1952	46.6		26.8	3.2		36	42	36111
Central Africa									
Algeria	3272	4.7					1	17	−340
Cameroon	2792	7.3	35.5	31.7	5.2		39	21	112096
Central African Republic	1720	61.8					36	8	13433
Central America									
Chad	2133	39.7		39.9	13		4	40	25387
Congo Rep.	2078	37.5		21.2	8.2	37.0	65	31	13303
El Salvador	2635	10.3	11.7	13.6	2.1		13	76	733

Gabon	2830	9.4					90	20	−94129
Guatemala	2478	15.8		46.5	0.7	60.8	33	36	8160
Honduras	2644	15.3					40	29	28116
Panama	2786	9.2					62	30	8563
Sudan	2336	25.2		38.2	16.3				
Oceania/Australia									
Fiji	2937	4.4					56	23	−2992
Timor-Leste	2131	27.2		50.2	11		45	26	6102

How GHI Scores Are Calculated

GHI scores are calculated using a three-step process.

First, values for each of the four component indicators are determined from the available data for each country. The four indicators are:

The percentage of the population that is undernourished

The percentage of children under 5 years old who suffer from wasting (low weight for height)

The percentage of children under 5 years old who suffer from stunting (low height for age)

The percentage of children who die before the age of 5 (child mortality)

Second, each of the four component indicators is given a standardised score based on thresholds set slightly above the highest country-level values observed worldwide for that indicator between 1988 and 2013 [3]. For example, the highest value for undernourishment estimated in this period is 76.5%, so the threshold for standardisation was set a bit higher, at 80% [6]. In a given year, if a country has an undernourishment prevalence of 40%, its standardised undernourishment score for that year is 50. In other words, that country is approximately halfway between having no undernourishment and reaching the maximum observed levels.

Third, the standardised scores are aggregated to calculate the GHI score for each country. Undernourishment and child mortality each contribute one-third of the GHI score, while the child undernutrition indicators child wasting and child stunting each contribute one-sixth of the score. This calculation results in GHI scores on a 100-point scale where 0 is the best score (no hunger) and 100 the worst. In practice, neither of these extremes is reached. A value of 100 would signify that a country's undernourishment, child wasting, child stunting and child mortality levels each exactly meet the thresholds set slightly above the highest levels observed worldwide in recent decades. A value of 0 would mean that a country had no undernourished people in the population, no children younger than 5 who were wasted or stunted and no children who died before their fifth birthday.

The scale below shows the severity of hunger from low to extremely alarming associated with the range of possible GHI scores using the revised formula.

Standardise component indicators:

$$\text{Standardised PUN} = \text{PUN80} \times 100$$

$$\text{Standardised CWA} = \text{CWA30} \times 100$$

$$\text{Standardised CST} = \text{CST70} \times 100$$

$$\text{Standardised CM} = \text{CM35} \times 100$$

Aggregate component indicators:

$$\begin{aligned}
&13 \times \text{Standardised PUN} \\
&\quad + 16 \times \text{Standardised CWA} \\
&\quad + 16 \times \text{Standardised CST} \\
&\quad + 13 \times \text{Standardised CM} \\
&\quad = \text{GHI score}
\end{aligned}$$

Determine values for the component indicators:

PUN Proportion of the population that is undernourished (in%)
CWA Prevalence of wasting in children under 5 years old (in%)
CST prevalence of stunting in children under 5 years old (in%)
CM proportion of children dying before the age of 5 (in%)

Standardise component indicators:

$$\begin{aligned}
\text{Standardised PUN} &= \text{PUN} / 80 \times 100 \\
\text{Standardised CWA} &= \text{CWA} / 30 \times 100 \\
\text{Standardised CST} &= \text{CST} / 70 \times 100 \\
\text{Standardised CM} &= \text{CM} / 35 \times 100
\end{aligned}$$

Aggregate component indicators:

$$\begin{aligned}
&1/3 \times \text{Standardised PUN} \\
&\quad + 1/6 \times \text{Standardised CWA} \\
&\quad + 1/6 \times \text{Standardised CST} \\
&\quad + 1/3 \times \text{Standardised CM} \\
&\quad = \text{GHI score}
\end{aligned}$$

In the countries included in the GHI, the share of the population that is undernourished stands at 12.3% as of 2015–2017, down from 17.6% in 1999–2001. Of children under 5, 27.9% are stunted based on data from 2013 to 2017, down from 37.1% in 1998–2002, and 9.3% are wasted, down slightly from 9.7% in 1998–2002. Finally, the under-five mortality rate was 4.2% as of 2016, down from 8.1% in 2000 [6].

Hunger hotspots are in sub-Saharan Africa and South Asia. Five of the eight countries where the hunger situation is classified as "alarming" lie in Africa: Central African Republic, Chad, Zambia, Sierra Leone and Madagascar. In many South Asian countries such as Pakistan, India, Nepal, Bangladesh, Myanmar, Laos, Cambodia and Indonesia, the situation is also classified as serious. In other Asian countries, the Middle East, Latin America, Caribbean and Eastern Europe, the values are usually classed as "moderate" or "low". For some states, there is no reliable data available.

8.4 Ways and Means for Solution to Hunger

1. Governments must ensure safe and nutritious food for all.

The policymakers should promote more sustainable solutions in agriculture, food safety, public health, livestock health and managing food chain.

2. Agriculture and food producers need to adopt good practices.

It should be a practice to encourage farmers to adopt more sustainable farming practices and reduce usage of chemicals within the farming system. The necessary precaution should be taken to avoid leaching of chemicals from the industries to the farm lands. These chemicals can cause reproductive and developmental problems, damage the immune system, interfere with hormones and cause cancer. Further, heavy metals such as lead, cadmium and mercury cause neurological and kidney damage and contaminate the air, water and soil.

3. Business operators must make sure food is safe.

The food production industries should follow practices like Hazard Analysis and Critical Control Points (HACCP), Good Manufacturing Practices (GMP) and Good Laboratory Practices (GLP). It is necessary to avoid food-borne illness, and the population will be safer from various harmful bacterial and viral diseases.

4. All consumers have a right to safe, healthy and nutritious food: Consumers should be given appropriate information about the food which they consume. This will help to take informed decisions and avoid complexities caused due to food allergies.
5. Food safety is a shared responsibility: The respective regional governments, regional economic bodies, UN organisations, development agencies, trade organisations, consumer and producer groups, academic and research institutions and private sector entities should come together and work on issues that are prevalent at both regional and global levels, for example, the spread of antimicrobial-resistant bacteria (AMR).

Table 8.3 Smart food in eight countries in South and Southeast Asia

Cereals	Roots and tubers	Pulses	Fruits and vegetables	Nuts, seeds and spices
Amaranth	Elephant	Black gram	Chayote	Linseed
Buckwheat	Foot yam	Cow pea	Drumstick	Nepali
Finger millet	Fancy yam	Faba bean	Fenugreek	Butter
Foxtail millet	Purple yam	Grass pea	Indian	Tree
Ragi	Swamp taro	Horse gram	Gooseberry	Nepali
Grain	Staechy grains	Lentil	Jackfruit	Pepper
Amaranth	Potato	Mung bean	Pumpkin	Perilla
Proso millet	Taro	Rice bean	Roselle	Walnut
Quinoa		Soybean	Snake ground	
Sorghum			Wood apple	
Specialty				
Rice				
Tartary				
Buckwheat				

Source: FAO [7]

6. It has been emphasised to strengthen green economy for improved human well-being and social equity and best approach for hunger solution. Climate-smart agriculture (CSA) is required to achieve sustainable food security by using crop diversity. It is necessary to assess the situation at regional level, define the baseline and also propose the alternative development pathways. The adoption of climate resilient crop varieties should be taken up along with cost-benefit analysis. The smart food in South and Southeast Asia is presented in Table 8.3. The adoption of CSA by farmers should be analysed, and the problem during adoption of germplasm conservation practices defines indicators of CSA. The production system efficiency of CSA needs to be defined during normal weather and extreme weather variation.

8.5 World Initiative for Zero Hunger

In spite of attention paid to agriculture development and food security over the past decades, there are still about 800 million undernourished and 1 billion malnourished people in the world. Besides this about more than 1.4 billion people are suffering from obesity, and one-third of all food produced is wasted. The global population is expected to reach 9.7 billion by 2050. The increasing global population is in synchronisation with global food consumption. If the present trends in consumption patterns and food waste continue, it is expected that we need more than 50% increase in food production by 2050. But total agriculture land is in decreasing phase because it is being converted for other purpose, such as highways, housing and factories. This loss of land is a large issue for agriculture, because it will make more difficult to produce enough food and ultimately lead to food insecurity, ecosystem instability, environmental degradation, poverty and hunger-like issues. In order to overcome with such problem, in 2015, the international community adapted a set of 17 goals as part of a new global agenda on sustainable development for a better world free from poverty, malnutrition and hunger in a favourable environmental surrounding with good health.

Most of the developing countries have been in a process of taking initiation in implementing SDGs. For example, underdeveloped country like Ghana in collaboration with HillCity College has been executing sustainable development projects in identified areas. The targets of these projects are to integrate the SDGs in models of development and entrepreneurship project works and research by students and trainee in the Institute. In India also the government has initiated several projects like Pradhan MantriJan Dhan Yojana (PMJDY) to promote women empowerment, eradicating poverty, Health for All, rural standard of life and education facilities. The government has disbursed a cumulative amount of 25 billion USD. About 329 million beneficiaries have been taken the advantage of this project.

In 2019 July, for the first time, UN High-level Political Forum on Sustainable Development (HLPF) 16 Asia-Pacific countries will announce their voluntary national reviews (VNRs) on implementation at national level. The main motto of the Agenda, to be presented during the HLPF session, being organised by UN Economic and Social Council (ECOSOC), is to increase the efficacy of the implementation of SDGs by sharing of national experiences including successes, challenges and lessons learned, in order to promote SDG implementation, globally. The 16 countries which are going to be associated with this mega event are Cambodia, Fiji, Indonesia, Iraq, Kazakhstan, Kuwait, Mongolia, Nauru, Oman, Pakistan, Palau, the Philippines, Timor-Leste, Tonga and Vanuatu.

Asia and the Pacific are also on the track to meet Sustainable Development Goals. Over the past few decades, Asia-Pacific has taken the credit of achieving poverty eradication campaign successfully. A remarkable reduction in rural poverty, increasing the level of education, extending life expectancy and building fast-growing and resilient economies, has been observed. But, energy transformation in Asia-Pacific is still lacking. So, the progress in SDG is not satisfactory.

The SDG Index report for 2018 presents a revised and updated performance of countries' distance to achieving the SDGs (Table 8.4). Among the countries that adapted SDGs, Sweden is at the top to achieve 84. 5% success to achieving the targets envisaged for 2030. The Scandinavian neighbours Denmark, Norway and Finland were close in achieving the goal.

Table 8.4 The SDG index

Rank	Country	Score
1	Sweden	84.5
2	Denmark	83.9
3	Norway	82.3
4	Finland	81.0
5	Switzerland	80.9
6	Germany	80.5
7	Austria	79.1
8	Netherlands	78.9
9	Iceland	78.4
10	United Kingdom	78.1
11	France	77.9
12	Belgium	77.4
13	Canada	76.8
14	Ireland	76.7
15	Czech Republic	76.7
16	Luxemburg	76.7
17	Slovenia	76.6
18	Japan	75.0
19	Singapore	74.6
20	Australia	74.5

References

1. FAO (Food and Agriculture Organization of the United Nations) (2015) Food security indicators (Updated May 27, 2015). http://faostat3.fao.org/download/D/*/E. Accessed 27 May 2015
2. Wiesmann D, Biesalski HK, von Grebmer K et al (2015) Methodological review and revision of the global hunger index. ZEF working paper 139. Bonn: Zentrum für Entwicklungsforschung (Center for Development Research). ww.zef.de/fileadmin/webfiles/downloads/zef_wp/zef_wp_139.pdf
3. Wiesmann D (2004) An international nutrition index: concept and analyses of food insecurity and undernutrition at country levels, Development economics and policy series 39. Peter Lang, Frankfurt
4. Alkire S, Santos ME (2010) Multidimensional poverty index: 2010 data. Oxford Poverty and Human Development Initiative, University of Oxford, Oxford. www.ophi.org.uk/policy/multidimensional-poverty-index/
5. Wiesmann (2006) Global hunger index: a basis for cross-country comparisons. International Food Policy Research Institute, Washington, DC
6. Black RE, Victora CG, Walker SP et al (2013) Maternal and child under-nutrition and overweight in low-income and middle-income countries. Lancet 832(9890):427–451
7. Asia and Pacific Regional Overview of Food Security and Nutrition (2018) Progress towards the SDGs, www.fao.org

Index

B. K. Behera et al., *Move Towards Zero Hunger*,
https://doi.org/10.1007/978-981-32-9800-2

MIX
Papier aus verantwortungsvollen Quellen
Paper from responsible sources
FSC® C105338

If you have any concerns about our products,
you can contact us on
ProductSafety@springernature.com

In case Publisher is established outside the EU,
the EU authorized representative is:
Springer Nature Customer Service Center GmbH
Europaplatz 3, 69115 Heidelberg, Germany

Printed by Libri Plureos GmbH
in Hamburg, Germany